AF899284

Studies in the History of Mathematics and Physical Sciences

7

Editor
G. J. Toomer

Advisory Board
R. Boas P. Davis T. Hawkins
M. J. Klein A. E. Shapiro D. Whiteside

Studies in the History of Mathematics and Physical Sciences

Volume 1
A History of Ancient Mathematical Astronomy
By **O. Neugebauer**
ISBN 0-387-**06995**-X

Volume 2
A History of Numerical Analysis from the 16th through the 19th Century
By **H. H. Goldstine**
ISBN 0-387-**90277**-5

Volume 3
I. J. Bienaymé: Statistical Theory Anticipated
By **C. C. Heyde** and **E. Seneta**
ISBN 0-387-**90261**-9

Volume 4
The Tragicomical History of Thermodynamics, 1822-1854
By **C. Truesdell**
ISBN 0-387-**90403**-4

Volume 5
A History of the Calculus of Variations from the 17th through the 19th Century
By **H. H. Goldstine**
ISBN 0-387-**90521**-9

Volume 6
The Evolution of Dynamics: Vibration Theory from 1687 to 1742
By **J. Cannon** and **S. Dostrovsky**
ISBN 0-387-**90626**-6

Volume 7
The Prehistory of the Theory of Distributions
By **J. Lützen**
ISBN 0-387-**90647**-9

Volume 8
Zermelo's Axiom of Choice: Its Origins, Development, and Influence
By **G. H. Moore**
ISBN 0-387-**90670**-3

Jesper Lützen

The Prehistory of the Theory of Distributions

With 29 Illustrations

Springer-Verlag
New York Heidelberg Berlin

Jesper Lützen
Department of Mathematics
Odense University
Campusvej 55, DK-5230 Odense M
Denmark

AMS Subject Classifications (1980): 01-A60, 46-03, 46F99

Library of Congress Cataloging in Publication Data
Lützen, Jesper.
The prehistory of the theory of distributions.
(Studies in the history of mathematics and physical sciences; v. 7)
Bibliography: p.
Includes index.
1. Distributions, Theory of (Functional analysis) I. Title. II. Series: Studies in the history of mathematics and physical sciences; 7.
QA324.L87 515.7′82 82–727
AACR2

© 1982 by Springer-Verlag New York Inc.
Softcover reprint of the hardcover 1st edition 1982

All rights reserved. No part of this book may be translated or reproduced in any form without written permission from Springer-Verlag, 175 Fifth Avenue, New York, New York 10010, U.S.A.

9 8 7 6 5 4 3 2 1

ISBN-13:978-1-4613-9474-7 e-ISBN-13:978-1-4613-9472-3
DOI: 10.1007/978-1-4613-9472-3

Preface

I first learned the theory of distributions from Professor Ebbe Thue Poulsen in an undergraduate course at Aarhus University. Both his lectures and the textbook, *Topological Vector Spaces, Distributions and Kernels* by F. Treves, used in the course, opened my eyes to the beauty and abstract simplicity of the theory. However my incomplete study of many branches of classical analysis left me with the question: Why is the theory of distributions important? In my continued studies this question was gradually answered, but my growing interest in the history of mathematics caused me to alter my question to other questions such as: For what purpose, if any, was the theory of distributions originally created? Who invented distributions and when? I quickly found answers to the last two questions: distributions were invented by S. Sobolev and L. Schwartz around 1936 and 1950, respectively. Knowing this answer, however, only created a new question: Did Sobolev and Schwartz construct distributions from scratch or were there earlier trends and, if so, what were they? It is this question, concerning the prehistory of the theory of distributions, which I attempt to answer in this book.

Most of my research took place at the History of Science Department of Aarhus University. I wish to thank this department for its financial and intellectual support. I am especially grateful to Lektors Kirsti Andersen from the History of Science Department and Lars Mejlbo from the Mathematics Department, for their kindness, constructive criticism, and encouragement.

My appreciation also goes to the Mathematical Institute at Utrecht University and the Department of History of Science and Medicine of Yale University for their hospitality during the six months I spent at each of those institutions. The help and encouragement I received from Dr. Henk Bos, Dr. Steven Engelsman (Utrecht), and Professor Asger Aaboe (Yale) were invaluable.

I also wish to thank Professor L. Schwartz and Professor H. A. Tolhoek for the information they provided in my interviews with them. In addition, my thanks goes to Professor Thomas, Professor Duistermaat, Professor E. Thue Poulsen, and Lektor Stetkær for the fruitful discussions they conducted with me.

I also wish to express my gratitude to Lenore Feigenbaum of Yale, who corrected my most glaring errors in English, to Mette Dybdahl, who painstakingly typed the manuscript and translated Ljusternik and Visik [1959] from the Russian, and to Springer-Verlag for their generous editorial care.

Odense
December 1981

JESPER LÜTZEN

Contents

Introduction

1. The historian's basic questions, . . . , are: What was the past like? and, How did the present come to be? The second question . . . How did the present come to be? . . . is the central one in the history of mathematics. [Grabiner 1975.]

This second question can only be answered if historians of mathematics follow mathematical developments up until the present day. Nevertheless, although the importance of the history of recent mathematics has been underscored by several mathematicians and historians, this portion of the history remains mostly uncultivated. In this book I try to cultivate a small, but what I consider important, corner of this field.

In 1900 Volterra called the nineteenth century the century of the theory of functions. F. E. Browder recently stated [1975]: "It would be equally appropriate to call the twentieth century the century of functional analysis." Thus the history of functional analysis is a central topic in the history of recent mathematics. Part of the history of functional analysis has fortunately been studied in some depth (see Ch. 1, §1). However, there still seems to be disagreement concerning the forces underlying the development of functional analysis. Some mathematicians hold the opinion that functional analysis emerged as a purely mathematical abstraction,[1] whereas J. Dieudonné [1975, p. 587], for example, has asserted "we never lost sight of the applications". In much of the discussion of the applicability of functional analysis the theory of distributions occupies a very essential position. Dieudonné [1964, p. 241] stated:

> The phenomenal growth of the theory of partial differential equations, during the last 10 years can also be taken as an excellent example of the impact of the general theory of topological vector spaces on classical analysis. Here the catalyst undoubtedly was the theory of distributions, although much of the technique is of earlier origin.[2]

F. E. Browder [1975] echoed this:

> In considering the applications of functional analysis in partial differential equations and in Fourier series analysis the theory of distributions stands out as an important and curious turning point.

However, this important turning point has not been studied in any detail from an historical point of view. My aim is to supply this deficiency.

2. The prehistory of the theory of distributions may also provide valuable material for philosophers of mathematics. In his monograph, *Bourbaki. Towards a Philosophy of Modern Mathematics*, Vol. I [1970], J. Fang pointed out how important the prehistory of the theory of distributions was from a philosophical point of view. His concluding words were [1970, p. 135]:

> *Philosophy* of mathematics should no doubt grapple with such a "product of human mind" [mathematics] and, as such, ought to examine the *modus vivendi* of working mathematicians and the *modus operandi* of their products, mathematics. But what should it be if it would hope to penetrate into the core of such problems? This question can hardly be answered since we have barely begun to formulate the question itself.

In a footnote he added:

> More specifically we may begin with the manner L. Schwartz and others founded the theory of distribution, for instance, which could rigorously and elegantly rationalize Dirac's delta-function, presenting in the process a new prospect to Fourier integrals and partial differential equations.[3]

However, I have not drawn very general philosophical conclusions from the history as told in this book, since I suspect that the development of the theory of distributions may not be representative of the way mathematics has developed in the twentieth century (cf. Concluding Remarks).

3. Although the theory of distributions in modern textbooks is presented in close connection with functional analysis, and although the prehistory of that theory adds a very important element to the history of functional analysis, the main trends in this prehistory are not to be found in functional analysis, but in different parts of concrete analysis and mathematical physics. This point of view was already put forth implicitly in the historical introduction to the first monograph on the theory of distributions: L. Schwartz' *Théorie des distributions* [1950/51]. Schwartz wrote nothing on the history of functional analysis, but gave a series of examples of problems and theories which were clarified by the theory of distributions and which had in turn anticipated the theory of distributions. He concluded:

> Nous voudrions avoir montré par ces exemples que la théorie des distributions n'est pas absolument une "nouveauté révolutionnaire". Beaucoup de lecteurs y retrouveront des idées qui leur étaient familière. Cette théorie englobe, de façon à la fois simple et correcte des procédés très hétérogènes et souvent incorrects utilisés dans des domaines très divers; c'est une synthèse et une simplification.

De Jager was of the same opinion [1964]:

> In the years between 1945 and 1949, L. Schwartz developed the theory of distributions by giving a synthesis, a generalization and a rigorous foundation of the work of many authors, who had already used the concept of distribution in a more or less cryptic way .... Among these there are ... mathematicians and physicists who were

> led to the use of distributions in connection with their investigations in applied mathematics and theoretical physics.

Dieudonné compared this aspect of the theory of distributions with the invention of the calculus [Dieudonné 1964, p. 241]:

> Schwartz' theory itself had had many forerunners, and indeed it may best be compared to what we call the "invention of the Calculus": it is quite clear that long before Newton and Leibniz, practically all prominent mathematicians of Europe around 1650 could solve most of the problems where elementary calculus is now used; but they had to resort to *ad hoc* considerations in each instance. Similarly most of the problems which belong to the theory of distributions had been considered and essentially solved before Schwartz, but no one had succeeded in building up a formalism which would dispense of special arguments in each particular case.

4. The problems or theories which shaped the prehistory of the theory of distributions are:

(1) Heaviside's operational calculus.
(2) Generalized derivatives and generalized solutions to differential equations.
(3) Generalized Fourier transforms.
(4) Improper functions; the δ-functions and the partie finie.
(5) De Rham's currents.

A discussion of the history of the last four theories and their connection to the theory of distributions constitutes the bulk of this book. For a treatment of the first theory the reader is referred to my paper, "Heaviside's operational calculus and the attempts to rigorize it" [Lützen 1979]. Just as this paper can be read in isolation, so too can Chs. 2–5 which contain accounts of the other theories, and each can be read separately. Each of these four chapters deals with a well-defined trend in the history of modern mathematics and may well be of interest also to readers who are not particularly interested in the theory of distributions.

I have chosen to end this *pre*history of the theory of distributions at the year 1950, for in that year Schwartz published Volume I of his *Théorie des Distributions*. This marked the point at which the theory of distributions began to be widely recognized as a mathematical *theory*. After 1950 other mathematicians suggested alternative definitions of generalized functions. These are of some interest to us since they shed light upon several arguments and theories prior to 1950. Although these definitions do not belong to the prehistory but to the history of the theory of distributions, I have briefly summarized them in an appendix, since they are not as well known as Schwartz' distributions.

This book is not primarily a story about how Sobolev or Schwartz developed the theory of distributions. All the important techniques and theories which anticipated the theory of distributions will be discussed, irrespective of their importance in the creative process which led these two mathematicians to their theories. The reader who is only interested in the work of these two people is referred to Ch. 2, §54–62 and Ch. 6, where Sobolev's and Schwartz' inventions are treated. These two sections can be read independently of the rest.

My account of the prehistory of the theory of distributions is based mainly on published sources. However, in a few cases information from unpublished material and interviews has been used as well.

Since so many individuals have contributed to the prehistory of the theory of distributions, space allows only short biographical sketches of the most important figures. Biographies of some of the mathematicians mentioned can be found in the *Dictionary of Scientific Biography* (*D.S.B.*). However, many have not satisfied the requirements for entry in the *D.S.B.*, either because they are still alive or because they have not been considered important enough. References to biographies of some of these mathematicians can be found in [May 1973].

5. The primary aim of the book is to show how different problems gave rise to theories anticipating the theory of distributions and how these theories were connected with each other and with the theory of distributions. A mathematically satisfactory description of these problems and theories would in many cases involve technical details which are of no interest in the present connection. In such cases, in order to accent the main ideas as clearly as possible, I have used a rather imprecise description of the uninteresting details. For example, a boundary curve or a function may be described as "sufficiently regular". Therefore even modern mathematical theorems in this book should never be accepted uncritically. On the other hand it is mainly addressed to mathematicians who possess some knowledge of the theory of distributions. To understand all of the mathematical arguments, the reader's knowledge of distributions should be on a level corresponding to the first two parts of F. Treves' *Topological Vector Spaces, Distributions and Kernels* [1967]. However, readers who are only familiar with the basic ideas of distribution theory need only omit a few insignificant details.[4] It is even my hope that some parts of this book will motivate students of mathematics to study the theory of distributions.

Modern mathematical terminology as found, for example, in Treves' book [1967] has been used without explanation. In places in which outmoded notation different from Treves' has been used in discussing early works, its meaning is explained. Throughout the book the word "classical" refers to the rigorous methods which were rooted in the last part of the nineteenth century. For the mathematics of this century "classical" is used synonymously with "nondistributional".

As a rule the term "generalized function" is used to describe any generalization of Dirichlet's concept of function, whereas "distribution" is used more specifically for objects equivalent to Schwartz' distributions. In some cases, however, a space of generalized functions is called a space of distributions even if it is only equivalent to one of Schwartz' spaces locally.

As usual a.e. means almost everywhere and a.a. means almost all in the Lebesgue sense.

6. The book is divided into six *chapters*. In the first the development of functional analysis is summarized. The next four chapters discuss the four main trends of the prehistory mentioned above. The last chapter deals with L. Schwartz' creation of the theory of distributions. Some of the chapters are divided into shorter *parts*. All chapters are subdivided into *sections*. In each chapter the formulas are numbered. When referring to sections, the number is preceded by a §, e.g. (Ch. 2, §10). A reference to a formula appears as (Ch. 2, (10)). When I refer to a section or a formula in the same chapter, I omit the number of the chapter, e.g. (§10) or (10).

The formulas (and references to formulas) in the quotations are renumbered to fit into the consecutive numbering of the other formulas in each chapter. These alterations have not been noted explicitly.

Reference to works mentioned in the bibliography are given by author's name and year of publication in square brackets. In rare instances the year refers to the year of composition. If the bibliography contains several publications by one author from the same year these are labelled *a*, *b*, *c*, . . . , e.g. [Schwartz 1947a]. In cases where it is clear which author is cited, only the year is given in square brackets. As a rule, all references included in parentheses refer to places in this book and all those in square brackets refer to other publications mentioned in the bibliography.

Year of birth and death of most of the persons mentioned in this book can be found in the index. In many cases I have been unable to find one or both of these years. Thus, when only the year of birth is given this does not necessarily mean that the person in question is still alive.

Chapter 1

Distributions in the Development of Functional Analysis

In this chapter I shall first sketch briefly the history of functional analysis in the first half of the twentieth century. Secondly, I shall point out more specifically certain theorems and theories in functional analysis and its applications which anticipated the theory of distributions.

1. The following account of the development of functional analysis will be very brief. More comprehensive treatment can be found in [Monna 1973], [Kline 1972, Ch. 46], [Dieudonné 1978, Vol. II], [Bernkopf 1966], and [Bourbaki 1969], on which this summary is based.

The motivation for the development of functional analysis came from two branches of classical analysis: the calculus of variations and the theory of integral equations. The first, which originated in the Italian school of variational calculus, reached its peak with Fréchet's (born 1878) doctoral thesis of [1906]. In what he called functional calculus, Fréchet initiated the study of abstract function spaces. He treated sets of functions in which a concept of limit was defined. On such spaces he considered functionals, i.e. real-valued functions defined on the function space. A functional could typically be a variational integral, i.e. the integral the maximum or minimum of which is sought. In that case, the function space would be the set of admissible functions. In his study of the calculus of variations Volterra had already introduced functionals in 1887 under the name "functions of lines". J. Hadamard had renamed them "functionals" in 1903. The main aim of Fréchet and his predecessors was to find a suitable definition of the differential of a functional so that the functional (e.g. the variational integral) would have an extremum at the points where the differential vanished. However, Fréchet did not quite succeed in setting forth such a theory. Neither were his function spaces and functionals applicable as a tool outside of the specific domain of the calculus of variations. They were too general, primarily because they had no linear structure.

2. Linear function spaces emerged from work on integral equations. In this domain David Hilbert's (1862–1943) *Grundzüge einer Allgemeinen Theorie der Linearen Integralgleichungen* [1912] was a highlight. The book summarized his achievements for the period 1904–1910, which in turn were motivated by the work of I. Fredholm. At the basis of Hilbert's discussion was the equivalence of integral equations and infinite systems of algebraic equations in infinitely many unknowns. In one case he obtained this equivalence by dividing the interval, in which the solution is sought, into an increasing number of equidistant subintervals. In another case the equivalence was obtained by expanding the functions with respect to what we would call a complete orthonormal system of functions. In both cases the square integrable functions and the related square summable sequences came to play a central role in Hilbert's approach to integral equations. However, even though he possessed the requisite technical apparatus, he never considered L^2 or l^2 as spaces in which geometric intuition and geometric notions could be used.

Erhard Schmidt undertook the geometrization of l^2 in [1908], thereby giving the first example of a Hilbert space. By then another example L^2, of a Hilbert space was easily at hand since F. Riesz and E. Fischer the previous year had shown that L^2 was in a 1–1 correspondence with l^2 [Fischer 1907, Riesz 1907]. Hilbert spaces and the theory of integral equations proved to be of value to the new quantum mechanics developing in the 1920s. In order to provide a rigorous foundation for this branch of modern physics, J. von Neumann (1903–1957) in [1927] offered an axiomatic approach to separable Hilbert spaces, thereby embedding both L^2 and l^2 in a more general structure.

3. However, Hilbert spaces were not the first linear function spaces to be axiomatized. An axiomatic treatment of the more general normed spaces (e.g. Banach spaces) had already been given during the years 1920/22 by Stefan Banach (1892–1945) among others. The L^p spaces were the first examples of Banach spaces where the norm was not defined by an inner product. These spaces were defined and discussed in detail by F. Riesz (1880–1956) in [1910] in connection with integral equations and the so-called moment problem. The abstract axiomatic treatment of normed spaces reached its high point in 1932 with Banach's book *Théorie des Opérations Linéaires* [Banach 1932], which, more than any other publication, helped to make functional analysis a separate mathematical discipline.

Only minor advances in the theory of normed spaces were made during the time between 1932 and the 1960s, when renewed interest in the field led to many deep results, for example, about the relationship between the analytical and geometrical aspects of Banach spaces. The death of many of Banach's collaborators during the Second World War gives a partial explanation of the 30 years of stand-still in the development of Banach space theory. More explanation can be found in the fact that Banach had developed his theory to such an advanced level that many of his famous open problems could not be solved until other branches of functional analysis had been explored. Moreover, the theory suffered from the lack of new applications. Many of the interesting function spaces as, for example, the holomorphic functions $H(\Omega)$ and the infinitely often differentiable functions $\mathscr{E}(C^\infty(\Omega))$ in a domain Ω, are not normable. Some of these more general topological vector spaces were studied to some extent in Banach's book, but a really fruitful step was not taken until 1935, when von Neumann gave the axiomatic definition of the locally convex spaces [von Neumann 1935].

4. The two objects of primary interest in the theory concerning the abstract spaces mentioned above were operators and functionals. Operator theory grew naturally out of Hilbert's theory of integral operators. The proof of the spectral theorem under weaker and weaker assumptions was the central problem in this branch of functional analysis. However, since Murray and von Neumann's joint papers of 1936–1943, *spaces* of operators have been at the heart of operator theory instead of the study of single operators.

From the point of view of the theory of distributions, the trend of functional analysis dealing with functionals is more interesting. As mentioned in §1 non-linear functionals had been considered in the early period of functional analysis. The first interesting results in the theory of continuous *linear* functionals were a group of representation theorems, the first of which was discovered in [1903] by Hadamard. He found that the continuous linear functionals T on $C(I)$ could be characterized as limits of sequences of integrals:

$$T(f) = \lim_{n\to\infty} \int_I f(t)g_n(t)\,dt, \tag{1}$$

where g_n are continuous functions depending only on T. F. Riesz improved Hadamard's theorem in [1909] showing that a continuous linear functional T on $C(I)$ could be represented as a Stieltjes integral (introduced by Stieltjes [1894])

$$T(f) = \int_I f(t)\,d\alpha, \tag{2}$$

where α is a function of bounded variation depending only on T. The third representation formula was found by F. Riesz in the following year [1910]. It stated that continuous linear functionals T on $L^p(I)$ could be written as

$$T(f) = \int_I f(x)g(x)\,dx, \tag{3}$$

where g is in $L^q(I)$ $(1/p + 1/q = 1)$.

Riesz's second theorem showed that some functionals could be represented by ordinary functions, in this case belonging to a different function space (unless $p = 2$). Later the representation of functionals by means of the integral (3) was to become the central tool in embedding the locally integrable functions L^1_{loc} into the space of distributions $\mathscr{D}'$. Riesz' first representation theorem, on the other hand, showed that not all functionals could be represented in the form of (3). Therefore it was implicitly known from 1910 that dual spaces—spaces of continuous linear functionals—offered the possibility for a generalization of the function concept. After F. Riesz had revived interest in the Stieltjes integral, Radon [1913] showed how these integrals could be defined according to Lebesgue's procedure using a measure (an additive set function) different from the Lebesgue measure.

5. Thus by 1913 the technical tools required for a generalization of the function concept in terms of measures had been discovered. However, neither functionals nor measures were used for this purpose until 1936 when Sobolev began to use functionals as generalized functions in his work on partial differential equations (however, see Wiener's remarks, Ch. 3, §7). There were two reasons why the theory of generalized functions required so much time to develop: the absence of strong motivation and the lack of profound understanding of the theory of dual spaces.

It will be seen in the subsequent chapters that motivation for a generalization of the function concept existed in the second decade of the twentieth century in many branches of mathematics, but the anticipatory ideas were too scattered to give rise to immediate action. Moreover the δ-function, whose nonrigorous foundation called attention to the desirability of a theory of generalized functions, was treated mainly by engineers and later by physicists who knew no functional analysis.

The theory of duality had to be more fully understood and absorbed before mathematicians could use it for a purpose so different from that for which it had been created. This occurred in the decades after Riesz' discovery of his representation theorems. Thus the Hahn–Banach theorem, on the extendability of a continuous linear functional, was proved by Hahn and Banach independently in 1927 and 1929 and the theory of duals of normed spaces was treated in detail in Banach's book [1932]. The theory of duality in more general locally convex spaces was studied by G. W. Mackey in [1943] and [1946].

L. Schwartz also worked with general duality theory during the Second World War. He abstractly treated the dual of $C^\infty(\mathbb{R})$ (or $\mathscr{E}$) which in his later theory was to play the role of the space of distributions with compact support. However, at the time he developed the abstract theory of duality he did not extend the operations of ordinary analysis to the new space of functionals. With this he could have generalized the concept of functions. Not until he had been motivated by concrete analysis did he see that functionals provided a basis for such a generalization (for more details see Ch. 6).

This shows clearly that the knowledge of the abstract theory of duality was not a sufficient condition for the creation of the theory of distributions. Was functional analysis then a necessary condition? Schwartz' theory of distributions shows that it was not necessary to have the whole abstract apparatus at hand for the discovery of the theory of distributions. For example, the main space of functions C_c^∞ was a type that had not been investigated. Even so, Schwartz' theory drew heavily on ideas and theorems from the abstract theory. Sobolev's work, however, shows that distributions could be constructed and applied in limited cases with very little general theory involved. Yet, he was also inspired by the general ideas on duality which existed at the time.[1]

Could a theory of generalized functions have been developed without a theory of duality? Yes, it could have been, and in fact such theories were actually developed. Of course, they were different from the distributions of Sobolev and Schwartz. Two of these alternative theories were proposed by Tolhoek, but never published [Tolhoek 1949] (see Ch. 4, §45–48). A third alternative method for defining distributions was invented by L. Schwartz while he was in the process of creating the theory of distributions (see Ch. 6, §6). There can be no doubt that one of these three alternative theories would have prevailed if the advanced state of duality theory had not made Schwartz' theory of distributions possible.

Thus, even though the advanced state of functional analysis was instrumental in the creation of Schwartz' version of the theory of distributions, it was neither a necessary nor a sufficient condition for the creation of a theory of generalized functions. The problems and theories which motivated the theory of distributions and which would have given rise to a theory of generalized functions, even without the theory of duality, arose, not in functional analysis, but in concrete analysis and mathematical physics. Before I discuss these (Chs. 2–5), I shall briefly mention another application of the theory of duality to concrete analysis, namely Fantappié's analytic functionals, which, according to Schwartz [1950/51, Vol. I, p. 8] "procèdent d'idées analogues".

6. Luigi Fantappié (1901–1956) published a series of articles beginning in 1930 on what he termed "analytic functionals". Many of his ideas were incorporated in his monograph, *Teoria de Los Funcionales Analiticos y Sus Aplicaciones* [1943a], which is based on courses he gave on the subject in Madrid and Barcelona in 1942/43.[2] Fantappié considered the space of ultra-regular functions on the complex sphere[3] equipped with an appropriate topology.[4] He defined an analytic curve to be an ultra-regular function $y(t, \alpha)$ depending analytically on the complex parameter α. Using these basic notations he stated that a functional F was analytic if:

(1) It was defined on an open subset of the ultra-regular functions.
(2) $F(y) = F(y_0)$ if y is an extension of y_0.
(3) For an analytic curve $y(t, \alpha)$ the function:

$$F(y(t, \alpha)) = f(\alpha) \tag{4}$$

is analytic and regular at those points α for which $y(t, \alpha)$ belongs to the domain of F.

The function $y(t, \alpha) = 1/(\alpha - t)$ is a special analytic curve which Fantappié used to define the indicatrix

$$F\left(\frac{1}{\alpha - t}\right) = \gamma(\alpha) \tag{5}$$

of a functional F.[5] The indicatrix was very important because it helped Fantappié prove the following representation theorem:

$$F(y(t)) = \frac{1}{2\pi i} \int_C \gamma(t) y(t)\, dt, \tag{6}$$

where γ is the indicatrix of the functional F and C is an appropriately chosen contour.[6] The representation (6) of a functional as a Cauchy integral is the main theorem in Fantappié's theory of linear analytic functionals. By means of it he gave a new basis for the operational calculus [1943, Ch. 8].

7. To accomplish this, he interpreted an operator B:

$$f(z) = By(t) \tag{7}$$

as a "mixed" functional, i.e. a functional F depending on a parameter z:

$$f(z) = F(y(t), z). \tag{8}$$

One of the problems of the operational calculus is to define a function g of an operator B. Fantappié started out with the following analysis: if it is possible to define $g(B)$ in such a way that certain linearity and continuity conditions[7] are satisfied, then for fixed B and a fixed function f the functional F defined as

$$F(g(\lambda)) = g(B)f \tag{9}$$

is a linear analytic functional in g. Therefore, its indicatrix

$$F\left(\frac{1}{\alpha - \lambda}\right) = \gamma(\alpha) \tag{10}$$

exists. Fantappié showed that γ must satisfy the equation

$$\alpha\gamma - B\gamma = f. \tag{11}$$

Conversely if γ satisfies equation (11) and $g(B)f$ is *defined* by

$$g(B)f = \frac{1}{2\pi i} \int_C g(\lambda)\gamma(\lambda)\, dt,^8 \tag{12}$$

then, Fantappié showed, the operators $g(B)$ satisfy most of the ordinary algebraic rules for real functions.[9] Thus they can be used freely, as in the operational calculus.

Fantappié extended this technique to a set of mutually commuting operators and showed how it could be used to solve a Cauchy problem for a partial differential equation with constant coefficients and analytic right-hand side.[10]

8. Both Fantappiè's theory of analytic functionals and Schwartz' theory of distributions are examples of applications of the theory of functionals to concrete analysis. However, there are many fundamental differences between the two theories. Fantappié's goal was to study analytic functionals abstractly and to rigorize the operational calculus, whereas Schwartz' goal was to extend the function concept. Although Fantappié had a method for identifying functions γ with functionals F (6), just as Schwartz had (3), he did not give a method for generalizing the function concept, since all his analytic functionals were represented in this way.[11] While Schwartz was interested in investigating highly irregular "functions" Fantappié's quantities were all regular. However, after the theory of distributions had been discovered, it was shown how analytic functions could be used to define distributions and even more irregular generalized functions called hyperfunctions (see Appendix, §5 and Ch. 3, note 18).

Chapter 2

Generalized Differentiation and Generalized Solutions to Differential Equations

Introduction

1. In the following, generalized differentiation means "differentiation" of functions which are not differentiable; similarly, a generalized solution to an *n*th order (partial) differential equation means a "solution", in some sense, which is not *n* times differentiable.[1,2] I shall not cover all phenomena which fall under the head of this chapter, but only those ideas which are related to the theory of distributions. Thus I shall only deal with piecewise differentiable functions in connection with the vibrating string while differentiation almost everywhere will be treated in some detail. The operational calculus, which contains generalized differentiation, is treated in [Lützen 1979] and in Ch. 4, §21–26.

2. Schwartz writes in the historical introduction to the *Théorie des Distributions* [1950/51]:

> Il est ensuite nécessaire d'établir les règles de calcul sur les distributions de façon à concilier les règles usuelles du calcul différentiel et celles du calcul symbolique. Et avant tout, il faut introduire une bonne définition de la dérivée. Il est assez curieux que cette nouvelle définition ait été peu à peu introduite, tout à fait indépendamment [of the other parts of the prehistory], dans la théorie des équations aux dérivées partielles. On peut écrire l'expression générale d'une solution de l'équation aux dérivées partielles $\partial^2 U/\partial x^2 - \partial^2 U/\partial y^2 = 0$ sous la forme $U = f(x + y) + g(x - y)$; mais une telle fonction U ne peut vérifier l'équation aux dérivées partielles que si f et g sont deux fois dérivables. Dans le cas contraire, on peut convenir de dire que U est "solution généralisées" de l'équation. Des définitions générales de ces solutions généralisées ont été données par divers auteurs, assez indépendamment les uns des autres (elles coïncident avec notre définition quand la solution généralisée est une fonction): Leray (dans sa thèse, sur les solutions "turbulentes" des équations aux dérivées partielles), Hilbert–Courant, Bochner ("solutions faibles") et moi-même. Remarquons qu'on définit ainsi U comme solution généralisée de $\partial^2 U/\partial x^2 - \partial^2 U/\partial y^2 = 0$ sans donner pour cela un sens précis à $\partial^2 U/\partial x^2$ et à $\partial^2 U/\partial y^2$. Dans le

> même ordre d'idées, également à propos d'équations aux dérivées partielles, Soboleff, Friedrichs et, récemment, Kryloff ont étudié une "dérivée généralisée" d'une fonction (la définition est identique à la nôtre, mais limitée au cas où la dérivée généralisée de la fonction est elle-même une fonction).

Schwartz' survey indicates that this chapter in the prehistory of distributions must be considered the most important in many respects; for not only did Schwartz' original ideas about distributions develop in connection with generalized solutions to partial differential equations, but it was also in this connection that distributions were defined for the first time, namely by Sobolev in 1935. Only the chapter on the δ-function is comparable to the present one in importance.

The passage quoted above leaves many questions unanswered. Which problems in addition to the vibrating string motivated the introduction of generalized derivatives and solutions? Which mathematical ideas were used as a basis for the definitions? What occurred in the period between the heated debate on the vibrating string around 1760 and the discoveries around 1934–1950 which Schwartz mentions?

I shall try to answer these questions in this chapter.

3. First, however, I will give an overview of the developments from 1750 to 1950. This two-hundred-year period can be divided into three periods characterized by differences in the treatment of generalized derivatives and solutions. In the first period, from 1750 to about 1840, nondifferentiable functions were to a large degree considered as acceptable solutions to partial differential equations although there was much discussion about it. The following 60 years bore both positive and negative impress of the rigorization of analysis. Therefore, it is no surprise that differentiation was reserved for differentiable functions and generalized solutions to differential equations were "prohibited". However, extensions of the classical concepts were in such demand that in the third period, from 1900 to 1950, different generalizations of derivatives and solutions were made, this time on the basis of the rigorous methods established in the previous period.

A description of this development in Hegelian terms is easy to see

thesis 1750–1840 ——— antithesis 1840–1900

\ /

synthesis 1900–1950.

4. In the following account the emphasis will be on the third period while the second will obviously only get a superficial treatment. The material in this chapter has been divided according to the different problems which gave rise to the generalizations. Therefore the history of the mathematical methods underlying the generalizations (as, for example, differentiation almost everywhere, test functions and adjoint operators) does not stand out clearly enough. Hence, I have compiled the history of these methods in the last part of the chapter (§63–64).

The division of the chapter will be as follows. First an account of the discussions on the vibrating string (§5–13), and then the implementation of rigour (§14–17); two short parts on the interest in generalized differentiation which emerged from the fundamental theorem of the calculus (§18–21) and the variational calculus (§22–27); four sections on generalized solutions to partial differential equations: one each on elliptic (§28–41) and hyperbolic equations (§42–51); a special section on operators in Hilbert spaces (§51–53); and a concluding section on Sobolev's work (§54–62).

The two charts at the end of the book (pp. 222 and 223) show the most important figures in the development, with arrows indicating that one has been cited by the other.

Part 1. Early Period. The Vibrating String

5. In recent years the vibrating string controversy has been studied thoroughly in order to shed light on the concept of function used in the eighteenth century (see, for example, [Ravetz 1961], [Youschkevich 1976] and [Lützen 1978]). Here I shall approach the discussion from another angle, namely as the first example of the use of generalized solutions to partial differential equations. Recently the subject has been studied from this point of view by Demidov [talk given at the *XVth International Congress for History of Science in Edinburgh, 1977*]. Since I shall carry the discussion further from a slightly altered point of view my account of the history will necessarily include some of the material covered in Demidov's paper.[3]

For mathematics in the eighteenth century our definition of a generalized solution in §1 is in fact anachronistic since the definition of differentiability was not made clear at that time. How can one then write about generalized solutions in connection with the arguments concerning the vibrating string? One can choose two different procedures. Either one can use the anachronistic term in the discussion or one can seek an eighteenth-century parallel to the modern term and base the arguments on it. I shall do both. The first method is valuable since it makes it possible to point out and explain obscure points in the arguments and to connect them to the modern theories. In order to understand the points of view of the participants in the discussion and in order not to distort the history completely I will also use the second method.

6. Let us then try to find the parallel to the generalized solution.

The seventeenth-century mathematicians did not work with a whole hierarchy of function spaces, such as the n times differentiable functions ($n = 0, 1, 2, \ldots$). They had one class, namely the analytic expressions, for which the calculus was supposed to be generally valid. An analytic expression is any combination of constants and variables built up from the known mathematical operations, both algebraic, and transcendental including infinite sums, differentiations and integrations. (I have discussed this in more detail in [Lützen 1978].) Thus the "classical solutions" at that time were the analytic expressions by Euler called continuous functions, whereas the generalized solutions were other types of functions called discontinuous, which were pieced together from a finite (or infinite) number of analytic expressions changing from interval to interval (or from point to point!) [Euler 1763]. In the following we shall call the Eulerian concepts E-continuous and E-discontinuous, in order to distinguish them from the modern concepts due to Cauchy. These notions will be made more concrete in the succeeding discussion of the vibrating string.

7. In [1747] J. L. d'Alembert (1717–1783) showed that the displacement of the points on a vibrating string fixed at the endpoints can be described by the expression

$$y = f(x, t) = \psi(x + t) + \varphi(x - t), \tag{1}$$

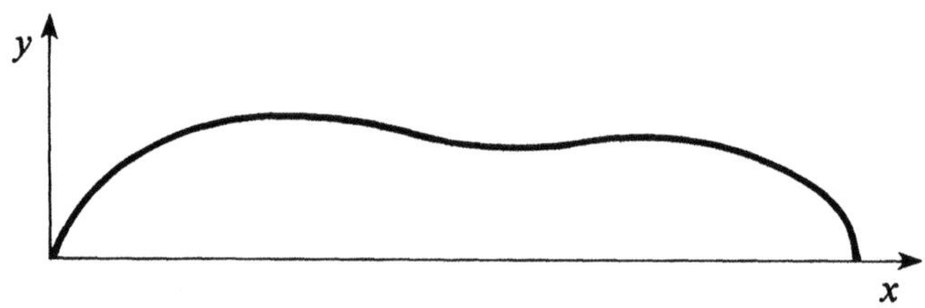

Figure 1

where the arbitrary functions φ and ψ can be determined from the initial state of the string. He did not explicitly set down the wave equation

$$\frac{\partial^2 f}{\partial x^2} = \frac{\partial^2 f}{\partial t^2} \tag{2}$$

which governs the movement, but that was done by L. Euler (1707–1783) in his second paper [1753] on the subject. In his first paper [1748] on the vibrating string, which he wrote immediately after the publication of d'Alembert's article, Euler derived expression (1) along the same lines as d'Alembert, i.e. by setting down equations in differentials equivalent to the wave equation (2). But his interpretation of the arbitrary functions φ and ψ differed decisively from d'Alembert's. D'Alembert had explicitly stated that φ and ψ must be analytic expressions whereas Euler thought that they could be completely arbitrary (E-discontinuous) with a hand-drawn curve as graph and could represent any initial state. It was this disagreement which caused the controversy.

8. First let us take a closer look at d'Alembert's opinion. D'Alembert had two sets of arguments. The first reflected a conservative attitude. He pointed out that Euler's use of the E-discontinuous functions was "contre toutes les règles de l'analyse" [d'Alembert 1761, p. 32] and that therefore a restriction to the analytic expressions was necessary. "Dans tout autre cas le problème ne pourra se résoudre au moins par les forces de l'analyse connue" [d'Alembert 1750, §2]. In this way he tried to convince mathematicians to use the calculus only within its then classical framework: the analytic expressions.

The other set of arguments which d'Alembert began to use in 1761 was based on an idea which in modern terms would state that the right-hand and left-hand second-order derivatives of the solution must be equal at every point. D'Alembert, however, did not express himself as directly as that; hence let us examine one of his arguments [d'Alembert 1761, §7]: For the second-order derivative of a function $f(z)$ d'Alembert used the geometric analogue of the expression

$$\frac{d^2f(z)}{dz^2} = \frac{f(z) + f(z + 2\varepsilon) - 2f(z + \varepsilon)}{\varepsilon^2}, \tag{3}$$

where ε is an infinitely small positive quantity. Using this formula in expression (1) he obtained (again in geometric form)

$$\frac{\partial^2 f(x, t)}{\partial x^2} = \frac{\psi(x + t) + \psi(x + 2\varepsilon + t) - 2\psi(x + \varepsilon + t)}{\varepsilon^2} + \frac{\varphi(x - t) + \varphi(x + 2\varepsilon - t) - 2\varphi(x + \varepsilon - t)}{\varepsilon^2} \tag{4}$$

and

$$\frac{\partial^2 f(x, t)}{\partial t^2} = \frac{\psi(x + t) + \psi(x + t + 2\varepsilon) - 2\psi(x + t + \varepsilon)}{\varepsilon^2} + \frac{\varphi(x - t) + \varphi(x - t - 2\varepsilon) - 2\varphi(x - t - \varepsilon)}{\varepsilon^2}. \tag{5}$$

The two fractions involving ψ are equal, but, as d'Alembert remarked, in the two fractions involving φ, ε differed in sign. Therefore the wave equation

$$\frac{\partial^2 f}{\partial t^2} = \frac{\partial^2 f}{\partial x^2} \tag{2}$$

could not be satisfied unless these two terms were equal which would mean that the radius of curvature on both sides of any value of $(x - t)$ had to coincide, i.e. that the radius of curvature could not jump (avoir un saut). Since such jumps were allowed in Euler's solution, it had to be rejected.

A modern reader can easily see that d'Alembert's calculation does not really support the restriction of solutions to the analytic expressions but is a strong argument in favour of the modern concept of solution ($\phi \in C^2(\mathbb{R})$). As pointed out by Demidov [1977, p. 38], d'Alembert came close to such an insight in the last years of his life. Thus in the paper, "Sur les fonctions discontinues" [1780], he first treated the equations

$$\frac{\partial z}{\partial x} = \pm \frac{a\,\partial z}{\partial y}. \tag{6}$$

He required of a solution z that (6) be fulfilled in all of the following four cases:

$$\begin{array}{ll} dx \text{ positive}, & dy \text{ positive}, \\ dx \text{ positive}, & dy \text{ negative}, \\ dx \text{ negative}, & dy \text{ positive}, \\ dx \text{ negative}, & dy \text{ negative}. \end{array} \tag{7}$$

He generalized to higher-order equations:

> Soit l'équation de l'ordre n et $\varphi(x, y, \ldots)$ la fonction discontinue qui entre dans l'integrale et qui devient successivement $\Delta(x, y, \ldots)$, $\Xi(x, y, \ldots)$ la fonction discontinue ne pourra entrer dans l'intégrale que dans le cas ou pour toutes lès valeurs possibles de z l'équation différentielle aura rigoureusement lieux.

By this rather obvious statement he no doubt meant that if, for example, the solution $z = \varphi(x - y)$ changed from one expression φ to another expression ψ in the point (the line) $x - y = \xi = A$, then the following equation must hold:

$$\frac{\partial^k \varphi(\xi)}{\partial \xi^k} = \frac{\partial^k \psi(\xi)}{\partial \xi^k} \quad \text{for } \xi = (x - y) = A \quad \text{and for} \quad k = 0, 1, \ldots, n.^4 \tag{8}$$

Here and in another argument in [1780][5] d'Alembert came very close to the modern characterization of a classical solution. He deviated though from the modern formulation in not possessing a clear definition of the partial derivatives and in focusing on functions composed of two (or finitely many) analytic expressions.

9. Naturally Euler had to answer d'Alembert's criticisms. For the first set of arguments Euler could only agree that the calculus "comme elle a été traitée jusqu'ici ne sauroit être appliquée qu'à des courbes, dont la nature peut être renfermée dans une équation analytique" [Euler 1765a, §7]. But in contrast to d'Alembert, Euler was willing to attempt to extend the calculus to include the E-discontinuous functions:

> . . . il faudra avouer que cette recherche [of the vibrating string] nous ouvre une nouvelle carriere dans l'analyse, en nous mettant en état d'appliquer le calcul à des courbes, qui ne sont assujetties à aucune loi de continuité, et si cela a paru impossible jusqu'ici, la découverte sera d'autant plus importante. [Euler 1765a, §8.]

Euler found it not only desirable but necessary to make such an extension of the calculus since he felt that the E-discontinuous functions were *forced* on mathematicians as "constants" of integration in the theory of differential equations involving functions of two or more variables. Euler himself did very little to realize such a program, but he repeatedly encouraged mathematicians to take up the problem:

> Cette partie [of the calculus] dont nous ne connoissons presque encore que les premiers élémens mérite sans doute que tous les Géometres réunissent leurs forces pour la cultiver. [Euler 1765a, §32.]

10. However, in his attempt to meet d'Alembert's criticisms of the second kind, Euler did come up with some interesting explanations. He gave three types of counterarguments [1765b].

a. First he remarked [1765b, §44] that the geometrical construction of the movement of the string from the initial displacement and velocity was, also in the E-discontinuous case, "tirée de l'équation intégrale $y = \Gamma(x + ct) + \Delta(x - ct)$ qui renferme la solution du problème" and "c'est toujours des équations intégrales qui fournissent les constructions". In this remark Euler anticipated the technique most applied for generalizing partial differential equations, namely, the replacement of the original equation with another, in this case with a functional equation. Yet when this method was applied in the twentieth century the substitutions were different types of integral equations, and not explicit expressions for the general solution as was the case with Euler.

b. But Euler was not satisfied with this argument. He wanted to prove that the general solution (1) actually satisfied the wave equation. He considered a curve made up of two analytic expressions (parabolas) so that the radius of curvature made a jump where the two met (Figure 2). In relation to d'Alembert's argument that

$$\frac{\partial^2 f}{\partial x^2} \neq \frac{\partial^2 f}{\partial t^2} \tag{9}$$

at the junction Euler argued [1765b]:

> Mais quoi qu'on y commette quelque erreur, cette erreur n'affectera qu'un seul élément, et sera par conséquent sans aucune conséquence, étant toujours infiniment petite.

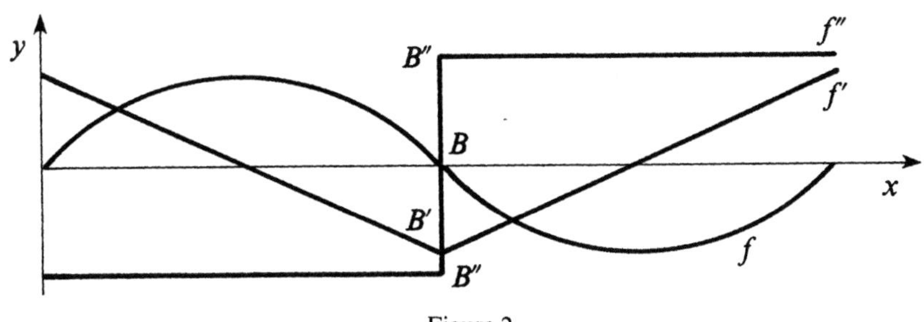

Figure 2

This argument clearly reflects the global ideas and the indifference to singularities which were characteristic of the calculus in the eighteenth century. It points towards the use of piecewise regular solutions to differential equations, a phase in the development I have chosen to omit.

c. Euler's last argument is the most interesting from the point of view of distribution theory. According to Euler, d'Alembert's counterexample worked because the second-order derivative had two values B'' and B'' at the junction (Figure 2). But, Euler argued [1765b, §48]:

> [on n'a qu'à émousser infiniment peu l'angle B' [the angle in the first derivative] dans la seconde ligne, pour réunir les deux points B'' et B'' en B et faire évanouir par ce moyen toutes les difficultés. Le chanchement qui en réjaillira sur la première courbe AMBma [the graph of f], ne sera aussi qu'infiniment petit, et partant ne changera rien dans l'état initial de la corde, d'ou la détermination du mouvement a été tirée.[6]

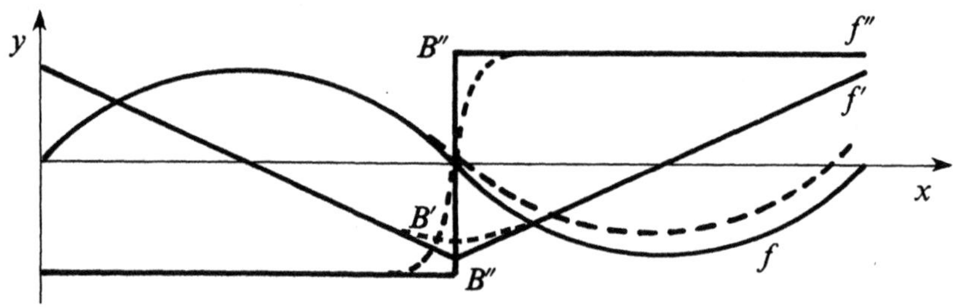

Figure 3

If, instead of the infinitely small alteration, we think of a sequence f_n tending to f (in some topology) then Euler's discussion can be rephrased: If a sequence of solutions $f_n(x \pm t)$ tends to f then f is a solution as well. If in the last sentence we read "generalized solution" instead of "solution" then we have a definition of a generalized solution to a partial differential equation: the sequence definition, which was proposed by several mathematicians in the twentieth century (see §63G).

Thus in the two last arguments (b) and (c) Euler tried to show that the error which d'Alembert had pointed out in Euler's solution was infinitely small, and he concluded:

> Toutes ces objections sont donc précisement de la même nature, que celle qu'on a faites autrefois contre le calcul différentiel, en lui reprochant que dans certains élémens quelques particules n'évanouissent point qu'on néglige néanmoins à l'égard des autres quantités. Comme aujourd'hui ces doutes sont entierement dissipés, ceux qu'on fait contre cette détermination du mouvement des cordes, tomberont aussi d'eux-mêmes.

To a modern reader, who is aware of the backward state of the fundations of analysis in the eighteenth century, this conclusion does not seem very convincing, nor was it to some of the contemporary mathematicians, among them d'Alembert. Moreover, in point of fact, Euler's arguments are wrong according to the standards of classical (i.e. nondistributional) analysis, whereas they can be given a correct meaning in the theory of distributions, as I have tried to do.[7]

11. With the paper, "Recherches sur la nature et la propagation du son" [1759], which brought the young mathematician J. L. Lagrange (1736–1813) his earliest success, he entered the controversy on the vibrating string. Lagrange agreed with d'Alembert that the solution of the wave equation must be an analytic expression:

> Il est certain que les principes du calcul différential et intégral dépendent de la considération des fonctions variables algébriques; il ne paraît donc pas qu'on puisse donner plus d'étendue aux conclusions tirées de ces principes que n'en comporte la nature même de ces fonctions. Or personne ne saurait douter que dans les fonction algébriques toutes leur différentes valeurs ne soient liées ensemble par la loi de continuité; c'est pourquoi il semble indubitable, que les conséquences, qui se déduisent par les règles du Calcul différentiel et intégral, seront toujours illégitimes dans tous les cas où cette loi n'est pas supposée avoir lieu. Il s'ensuit de là que, puisque la construction de M. Euler est déduite immédiatement de l'intégration de l'équation différentielle donnée, cette construction n'est applicable par sa propre nature qu'aux courbes continues. [Lagrange 1759, §15.]

But Lagrange thought that Euler's solution gave the correct description of the motion of the string. In order to show that this was the case he gave a new derivation of Euler's result, a derivation which in his opinion did not use differential or integral calculus. He first found the motion of a weightless string loaded with a finite number of point masses and then made the number of points tend to infinity. Thus Lagrange abandoned the differential equation and replaced it with a completely different mathematical description which he found more suitable for describing the physical reality.[8] An attitude very similar to Lagrange's was held by many of the mathematical physicists in the last part of the nineteenth century (see Appendix and §63A).

However Lagrange changed his mind. In his "Nouvelles recherches sur la nature et la propagation du son" [1760/61, in particular §5] he made the wave equation

$$\frac{\partial^2 z}{\partial t^2} = c\,\frac{\partial^2 z}{\partial x^2} \tag{10}$$

the basis of the analysis, just as Euler and d'Alembert had done. Multiplying with a function $M(x)$ and integrating partially over the interval $[0, a]$ he obtained

$$\int_0^a \frac{d^2 z}{dt^2} M\,dx = c\left(\frac{dz}{dx} M - z\,\frac{dM}{dx}\right)\Bigg|_0^a + c\int_0^a z\,\frac{d^2 M}{dx^2}\,dx.^{9} \tag{11}$$

Since he had assumed z to be zero at the endpoints $(0, a)$ of the string, the term $z(dM/dx)$ vanished. He considered only functions M with $M(0) = M(a) = 0$ so that (11) was transformed into the equation

$$\int_0^a \frac{d^2z}{dt^2} M\,dx = c \int_0^a z \frac{d^2M}{dx^2}\,dx. \tag{12}$$

Lagrange then proceeded to solve (12) using rather strange arguments with trigonometric integrals; what is more interesting to us however are his remarks on the passage from (10) to (12). He explained in [§8] how, by his method, the partial integration had provided him with an equation (12) in which "les différentielles de z dependantes de x s'évanouissent". Therefore:

> on n'a point à craindre d'introduire par là dans notre calcul aucune loi de continuité entre les différentes valeurs de z.

The resulting formula (12), however, still contained differentiation of z with respect to t, but Lagrange tried to convince the reader that the t dependence can be handled as in ordinary calculus:

> Cette intégration ne regarde que la variabilité de t, et elle s'achève selon les méthodes connues du calcul intégral, puisque ici la loi de continuité a lieu.

As Lagrange would easily have been able to see this does not make sense, because from the final expression

$$z = \varphi(x + \sqrt{c}t) - \psi(x - \sqrt{c}t) \tag{13}$$

of the solution it is obvious that z is equally as discontinuous a function of t as it is of x. But this flaw must not hide the fact that Lagrange discovered the main ideas of the test function generalization of a partial differential equation: the use of partial integration to remove the differential operator from the unknown function onto the test function M, which for simplicity is taken to vanish at the boundaries of the domain considered.

This method of generalization, which is at the heart of distribution theory, was first rediscovered by Wiener [1926b] (see §43–44) after having been used implicitly in variational calculus (see note 43).

With this new attitude towards the wave equation Lagrange had made himself vulnerable to d'Alembert's criticisms (§8). However he rejected the criticisms with an interesting argument [Lagrange 1760/61]: d'Alembert had calculated the second derivatives from the values of the function at the points $(t; t + dt; t + 2\,dt)$ and $(x; x + dx; x + 2\,dx)$, (4) and (5). Lagrange thought that this was a mistake and that instead one ought to use the points $(t - dt; t; t + dt)$ and $(x - dx; x; x + dx)$, i.e. that (3) ought to be replaced with:

$$\frac{d^2f(x)}{dx^2} = \frac{f(x - \varepsilon) + f(x + \varepsilon) - 2f(x)}{\varepsilon^2}; \qquad \varepsilon \text{ infinitely small.} \tag{14}$$

In this way Lagrange found that the expressions corresponding to (4) and (5) became identical also at points where the radius of curvature made a jump (but $f \in C^1$), the common value being the mean value between the right-hand and left-hand second-order derivatives. In this way the wave equation was reestablished.

The argument anticipates Riemann's consideration in [1854] (see §16). We can interpret (14) as a generalization of d'Alembert's formula (3) for the second-order derivative; Lagrange thought of it more as a correction than a generalization.

12. Thus the dispute about the vibrating string raised the question: How regular must a solution to a differential equation be? In addition to the answers we have already studied I shall briefly mention a few others.

Lagrange wavered and required first, that the function itself may not jump, and then, that the function together with all its derivatives may not jump [Lagrange 1764/65].

Marquis de Condorcet thought (1771) that a solution to nth order differential equations had to be so regular that the function itself and its first $(n - 1)$-derivatives could have no jumps. P. S. Laplace[10] came to the same result in [1772, *Oeuvres*, 10, p. 81] (repeated in [1812, *Oeuvres*, 7, pp. 77–80] and l'Abbé de Caluso did also in (1786/87) (see Arbogast [1791] for quotes from Condorcet and Caluso).

Already in his early works Gaspard Monge (1746–1818) took the same standpoint as Euler [Taton 1950, 1951, pp. 182–186]. He used a geometric argument representing a function of two variables by a surface in space. Monge showed in numerous examples how the surface representing the solution of a nth order partial differential equation in two variables could be constructed from n arbitrary curves E-continuous or E-discontinuous. Monge felt that the possibility of the construction proved that the arbitrary functions which appear in the integration of partial differential equations can be completely arbitrary.[11]

13. To get a clarification of the concept of a solution to a partial differential equation the St. Petersburg Academy posed the following prize problem:

> Déterminer si les fonctions arbitraires introduites par l'intégration des équations différentielles, qui ont plus de deux variables, appartiennent à des courbes ou surfaces quelconques, soit algébriques, transcendantes ou méchaniques, soit discontinues ou produites par le mouvement libre de la main; ou bien, si elle ne peuvent légitimement être rapportées qu'à des courbes continues et susceptibles d'être exprimées par des équations algébriques ou transcendantes. [Arbogast 1791.]

Louis Arbogast (1759–1803) won the prize in 1787 with a paper which supported Euler's point of view [Arbogast 1791]. He introduced the concept "discontiguous" to describe a function which has jumps in certain points, a property which, as we have seen, had been essential in many descriptions of the class of admissible solutions. Arbogast advocated the idea that arbitrary functions could be both E-discontinuous and discontiguous. He obtained his main arguments from Monge's geometric interpretation of the problem.[12]

The St. Petersburg prize essay marks the end of the controversy of the vibrating string, and Euler's ideas had come out victorious. But the prize essay was not *responsible* for the end of the dispute. The intensity of the discussion had already decreased by the early 1780s and the controversy would probably have died out by 1790 even without Arbogast's paper. One reason for the declining interest in these matters was that the two principal figures, Euler and d'Alembert, both died in 1783. Another reason could be that the younger participants had run out of arguments and had realized that with the available foundation for the calculus the problem could not be answered unambiguously.

Whatever ended the discussion, the result was that the subject matter became divided into two separate branches, one being the foundations of mathematics and the other the theory of partial differential equations. The rigorizing movement in analysis was started in the 1820s by A. L. Cauchy and culminated in the 1870s with a vast number of investigations carrying further ideas from K. Weierstrass' lectures. It was based on a broad analysis of the foundations, and the arbitrary functions played no part in the rigorizing process any longer.[13] As a result of this rigorization the concepts "n times differentiable" and "n times continuously differentiable" were clearly defined; furthermore, it became generally accepted among mathematicians that only differentiable functions could be differentiated, and thus solutions to nth order differential equations had to be n times differentiable.

Part 2. The Age of Rigour

14. The development of the theory of partial differential equations continued during the nineteenth century, but mathematicians turned to new and more fruitful problems: for example, the development of a theory of characteristics (Monge) and existence and uniqueness theorems for initial value problems with very regular initial values (Cauchy). The question of the regularity of the arbitrary functions was completely ignored. Even Monge, who had been interested in the problem in his younger days, used arbitrary functions in his famous *Application de l'Analyse à la Géometrie* [1807] without mentioning how nonregular such functions could be. One also looks in vain for differentiability conditions on the solutions in Cauchy's work on differential equations; even as late as E. J. B. Goursat's *Leçons sur l'Intégration des Equations aux Dérivées Partielles du Second Ordre* [1896/98] and his *Cours d'Analyse Mathématique* [1911] arbitrary functions are introduced without any mention of their regularity.

In spite of this neglect a classical notion of a solution to partial differential equations arose slowly and undramatically in the period 1840–1880. This can be seen in the papers on applied mathematics in which the problems of discontinuities could not be as easily ignored as they could be in pure mathematics. Thus in the theory of wave propagation the propagation of singularities attracted special interest.

The earlier problem of the propagation of a sharp bend on a vibrating string was taken up by E. Christoffel in "Untersuchungen über die mit Fortbestehen linearer partieller Differentialgleichungen verträglichen Unstetigkeiten" [1876] (see also Christoffel [1877]). He pointed out that at the singular point the wave equation could not be used; he also rejected the description of the movement of singularities in terms of Fourier series, a description already used by Fourier and Helmholtz [1865], because such a procedure was also derived from the wave equation. Instead he used the laws of collision and a so-called phoronomical equation, which expressed the coherence of the string, to derive a new set of equations governing the propagation of the singularity. He determined the movement of the regular portions of the string using the wave equation. A. Harnack used a slight variation of this method in "Ueber die mit Ecken behafteten Schwingungen gespannter Saiten" [1887].[14]

For the more difficult problem of the propagation of plane waves (sound waves) in air, Riemann (1826–1866) had already used a similar idea in [1858/59]. This problem was also taken up and clarified in Christoffel's article mentioned above [1876]. In this case a theory of propagation of singularities—shock waves—was even more unavoidable than it was for the vibrating string, for, as remarked by Riemann, even if the initial situation is described by very regular functions the propagating density waves will at some moment develop into shock waves (Figure 4).

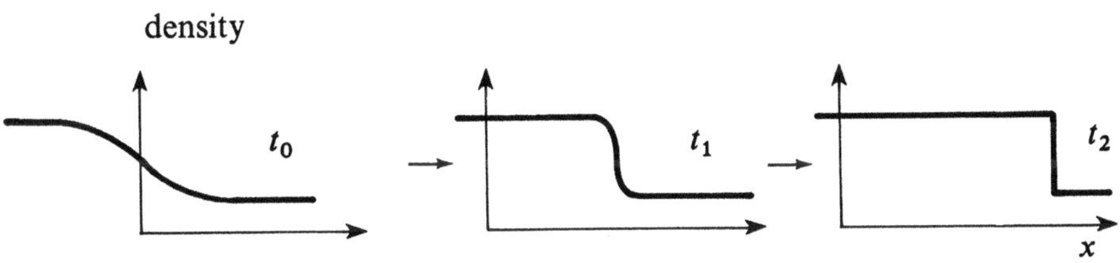

Figure 4

15. In conclusion it can be said of the nineteenth century that:

(1) Mathematicians did not explicitly comment on the regularity requirements which the classical definition of the derivative had imposed on the solutions to (partial) differential equations.

(2) When physical situations forced mathematicians to treat nonregular initial value conditions they replaced the differential equation (at the singular points) with other models of the physical system, so that generalized solutions were avoided. This method was in agreement with Lagrange's earliest ideas and can be found in textbooks from this century as well (cf. Riemann–Weber,

Die partiellen Differentialgleichungen der mathematischen Physik [1919, p. 210]).[15]

16. However, one can find two cases of generalized differentiation in the nineteenth century outside the framework of the theory of differential equations. One very conscious generalization was made by Dini; this will be put into context in the next section. Another was more implicitly proposed by Riemann in his Habilitationsarbeit on trigonometric series [1854]. He was interested in the behaviour of series of the form

$$\begin{aligned}\Omega = {} & a_1 \sin x + a_2 \sin 2x + \cdots \\ & + \tfrac{1}{2}b_0 \\ & + b_1 \cos x + b_2 \cos 2x + \cdots \end{aligned} \tag{15}$$

which were not *a priori* supposed to be Fourier series. He first considered the case in which the coefficients converged towards zero. Naturally Ω need not be convergent for that reason, but the twice term-by-term integrated series

$$\begin{aligned}F(x) = {} & C + C'x + b_0 x^2 \\ & - a_1 \sin x - \frac{a_2}{4} \sin 2x - \frac{a_3}{9} \sin 3x - \cdots \\ & - b_1 \cos x - \frac{b_2}{4} \sin 2x - \frac{b_3}{9} \sin 3x - \cdots \end{aligned} \tag{16}$$

is. Riemann's problem now was how to find Ω from F at the points where Ω converged. He could not use ordinary differentiation since F does not have to be twice differentiable (this is my consideration, not necessarily Riemann's). However Riemann could prove that the fraction

$$\frac{F(x + \alpha + \beta) - F(x + \alpha - \beta) - F(x - \alpha + \beta) + F(x - \alpha - \beta)}{4\alpha\beta} \tag{17}$$

for α, β tending to zero and the ratio α/β remaining finite would tend to Ω at the points where the sequence Ω converged.[16]

We see that Riemann's theorem is heavily dependent on the use of the generalized second-order derivative (17) which in the case $\alpha = \beta$ is identical to the generalization used by Lagrange [1760/61] (§11, (14)). However, as late as 1854 when Riemann wrote the above-mentioned paper, the now classical definition of the second-order derivative was not yet so "classical" that Riemann considered (17) to be a generalization; he called the limit of (17) the "zweite Differentialquotient" and not the generalized second-order derivative.[17,18]

17. This chapter has given the impression that the rigorizing movement in the nineteenth century prevented the discussion of generalized solutions to partial differential equations and also to a certain degree of generalized differentiation. In the long run, however, the rigorization of analysis had a positive influence on the development of the theory of distributions. A clear understanding of the classical concepts and theories was a necessary condition for development of the more general procedures which culminated in the theory of distributions. This is demonstrated clearly by the discussion of the vibrating string in which the many ideas which later proved to be fruitful could not demonstrate their force until they were seen in the light of the rigorous classical theory.

Until now we have only dealt with theories which, with our present knowledge, can be interpreted as generalizations of the classical concept of "differentiation" and "solution to a differential equation". The mathematicians discussed here either did not consider their theories as generalizations or they considered them to be generalizations of concepts or procedures differing from the classical ones.

In the following sections we shall discuss more conscious, explicit generalizations of the classical concepts.

Part 3. The Fundamental Theorem of the Calculus and the Determination of Areas of Surfaces

18. Differentiation of more general functions than the differentiable functions was treated in pure mathematics in connection with the fundamental theorems of the calculus. In their raw form these theorems assert:

(I) An *integral* $\int_a^x f(\xi)\,d\xi$ is *differentiable*, and

$$\frac{d}{dx}\int_a^x f(\xi)\,d\xi = f(x).$$

(II) A *derivative* $(d/d\xi)f(\xi)$ is *integrable*, and

$$\int_a^x \frac{d}{d\xi} f(\xi)\,d\xi = f(x) - f(a).$$

In these theorems it is of decisive importance how the terms *integral*, *derivative* and = are interpreted. It is obvious that a generalization of the concept of integral may destroy the validity of (I) whereas (II) may cease to be valid if the notion of derivative is generalized. On the other hand a generalization of the concept of differentiation can restore (I) and a broader definition of integration can restore (II). Thus from our point of view theorem (I) is the most interesting of the two, since the successive attempts to save it led to different generalized definitions of differentiation.[19]

19. Cauchy, who altered the main theorems (I) and (II) from *definitions* of the integral into real *theorems*, proved [1823a] their validity when *integrable* means continuous and a differentiable function is taken to mean a differentiable function with continuous derivative.[20] Thus Cauchy's version of the main theorems was nicely symmetric.

The symmetry was destroyed after Riemann's well-known extension of the integral in [1854] (published 1868),[21] for as Hankel showed in [1870], theorem (I) was invalid for the Riemann integral even if differentiable was taken to mean "differentiable everywhere" instead of "differentiable with continuous derivative".[22] A further extension was required.

Such a generalization of the derivative was given in 1878 by Ulyses Dini (1845–1918) who in his very influential book, *Fundamenti per la Teorica della Funzioni di Variabile Reali* [1878], defined what are now known as the Dini derivatives.[23] He showed that each of the four Dini derivatives of a Riemann integral $\int_a^x f(x)\,dx$ exists and differs from f by a function which has zero integral. This gave the desired version of theorem (I) when = was taken to mean = almost everywhere in Riemann measure.[24,25]

H. Lebesgue's (1875–1941) definition [1902] of a new integral was followed by another extension of the concept of differentiation to be used in (I), namely an extension to the functions differentiable a.e., in which case = in (I) must be understood as = a.e. Lebesgue showed [1904] that all continuous functions of bounded variation were differentiable a.e., but he found that a nice symmetrical version of the main theorems could be obtained by restricting the extension of differentiation to a subclass, namely to the functions which G. Vitali later [1905] gave the name absolutely continuous. Lebesgue stated that:

$\Updownarrow$ F is absolutely continuous in $[a, b]$

There exists an (L) integrable function f, such that (18)

$$F(x) - F(a) = \int_a^x f, \qquad \forall x \in [a, b]$$

and $f = F'$ a.e.

This theorem, which was proved by Vitali [1904/05] gave a unified symmetrical analogue to Cauchy's symmetrical version of the main theorems (I) and (II), but for a larger class of functions.[26]

20. Vitali [1907/08] extended theorem (18) to the plane after generalizing the concept of absolute continuity to two dimensions.[27] An alternative definition was given by L. Tonelli [1926a, b] in connection with investigations on the areas of surfaces. Tonelli wanted to characterize the surfaces given by

$$z = f(x, y), \qquad (x, y) \in Q \equiv [0, 1] \times [0, 1], \qquad f \in C_0 \tag{19}$$

for which the area, as defined for a very large class of surfaces by Lebesgue [1902], was given by the well-known formula

$$S = \iint_Q \sqrt{1 + p^2 + q^2}\,dx\,dy, \tag{20}$$

where

$$p = \frac{\partial f}{\partial x} \quad \text{and} \quad q = \frac{\partial f}{\partial y}.$$

Therefore he needed a condition which would guarantee the existence of the partial derivatives p and q in a generalized sense and make p and q sufficiently well behaved to ensure the existence of the integral (20). To do this he defined:

A continuous function f is of bounded variation in Q if

(1°) $f(\bar{x}, y)(f(x, \bar{y}))$ is of bounded variation in y, (x) for almost all fixed values of $\bar{x}$, $(\bar{y})$ in $[0, 1]$.
(2°) The total variation of $f(\bar{x}, y)$, $(f(x, \bar{y}))$ in $[0, 1]$ is integrable in $\bar{x}$, $(\bar{y}) \in [0, 1]$.

A continuous function f is absolutely continuous in Q if

(1°) $f(\bar{x}, y)$, $(f(x, \bar{y}))$ is an absolutely continuous function of y, (x) for almost all $\bar{x}$, $(\bar{y})$ in $[0, 1]$.
(2°) as (2°) above.[28]

With these new definitions at his disposal Tonelli could prove

(A) $\Updownarrow$ The surface described by (19) has finite Lebesgue area S, f is of bounded variation.

(B) If one of the equivalent statements of A holds true, then $\iint_Q \sqrt{1 + p^2 + q^2}$ exists and

$$S \geq \iint_Q \sqrt{1 + p^2 + q^2}.$$

(C) If f in (19) has finite Lebesgue measure the following equivalence holds:

$$\text{area } S = \iint_Q \sqrt{1 + p^2 + q^2} \Leftrightarrow f \text{ is absolutely continuous.}$$

In this way the question of the applicability of the formula (20) in the case of surfaces defined by (19) was completely settled. A treatment of surfaces given in parametric form was presented by C. B. Morrey [1933] among others, who used a third kind of absolute continuity.

21. The generalizations of differential operators to the spaces of absolutely continuous functions gave an early example of what are now called Sobolev spaces H_v^p.[29] In the two-dimensional case similar spaces were introduced both before and after 1926 (see §24 and §61) as a means of solving problems different from those of Tonelli and sometimes methods different from his were applied. As we shall see in the subsequent sections the connection between absolute continuity and other methods of generalizing differential operators was noted explicitly by several mathematicians. It is mainly these early comparisons that makes Part 3 interesting to us.

One problem in which spaces like the one defined above emerged, came from the variational calculus. Tonelli himself used his absolutely continuous functions in the calculus of variations in [1929], but Beppo Levi preceded him by more than 20 years.

Part 4. The Calculus of Variations

22. The motivation for introducing generalized derivatives into the calculus of variations was described clearly by L. C. Young [1938]:

> The simplest problem of the Calculus of Variations consists in determining the minimum of an integral
>
> $$\int_{x_0}^{x_1} f(x, y(x), y'(x))\, dx \tag{21}$$
>
> for a given $f(x, y, y')$ and fixed ends $y(x_0) = y_0$, $y(x_1) = y_1$ when $y(x)$ belongs to a certain class of "admissible curves" which must be explicitly defined, and which certainly includes all the functions $y(x)$ with continuous derivatives of all orders which fulfill the boundary conditions $y(x_0) = y_0, y(x_1) = y_1$.
>
> The classical theory of the Calculus of Variations was concerned with the case in which the admissible curves consisted of these analytic curves only. Modern researches have shown that in order to treat the classical problem satisfactorily it is convenient to enlarge the class of admissible curves. This is because the variational methods so far known depend on the existence of a "*minimizing*" *curve* in the class of admissible curves, i.e. a function $y(x)$ for which the minimum is attained. To ensure, as far as possible, that such a minimizing curve should exist, the class of admissible curves has, since the days of Weierstrass, been enlarged, successively, by the inclusion of functions $y(x)$ with bounded, piecewise continuous, derivative $y'(x)$ [Caratheodory 1904] by the inclusion of absolutely continuous functions $y(x)$ for which $y'(x)$ then denotes the derivative almost everywhere [Tonelli 1921]. by the inclusion of certain nonrectifiable curves [Menger 1936], and by the inclusion of "generalized curves". Another interesting extension, due to McShane [1933a] consists in the inclusion of certain admissible curves of a corresponding *parametric* problem.

There are two branches of this development which are of interest here. One related to variational problems in the plane, came from the newly introduced Lebesgue integral, and led to spaces similar to Tonelli's space of absolutely continuous functions (§23–25). The other gave rise to Young's generalized curves and surfaces, a generalization of a very different nature (§26).

23. The framework for the first generalization was the Dirichlet principle which deals with the solution of the Dirichlet *problem*:

> Given a function f on the boundary of a domain Ω. Determine a function u in $\overline{\Omega}$ such that
>
> $$\Delta u = 0 \text{ on } \Omega \quad \text{and} \quad u = f \text{ on } \partial\Omega.$$

The Dirichlet *principle* states that the Dirichlet problem has a unique solution, namely the function ($\in C^1(\Omega) \cap C^0(\bar{\Omega})$), which fulfills the boundary condition and minimizes the Dirichlet integral:

$$\int_\Omega \left[\left(\frac{\partial u}{\partial x}\right)^2 + \left(\frac{\partial u}{\partial y}\right)^2\right] dx\, dy. \tag{22}$$

The Dirichlet principle had been "proved" by Sir W. Thomson (later Lord Kelvin (1824–1907)) [1847] and L. Dirichlet (1805–1859) [1876][30] invalidated by K. Weierstrass (1815–1897), but reestablished under more restrictive conditions by D. Hilbert (1862–1943) [1905]. The difficulty with the principle was the existence of the minimum, whereas it is correct that if the minimum exists then it will be the solution to the Dirichlet problem, provided it is a C^2 function.[31,32]

Hilbert's proof gave the first example of the use of the direct method in variational calculus by which a minimum is found, not by solving the Euler equation, but by finding a sequence of functions which converges to the minimum. After Hilbert the method was simplified and extended to a larger class of boundary curves and values.[33]

24. The first such extension was given in [1906] by the Italian mathematician Beppo Levi (born 1875). Before this the class of admissible functions for the variational integral (22) had been taken to be $C^1(\Omega) \cap C(\bar{\Omega})$ for the integral the Riemann integral. Instead Levi used the Lebesgue integral, to which he was led not by "amore di generalità" (love for generality) but by "necessità di cose" (necessity). The generalization of the integral allowed him to use a larger class of admissible functions because, as he remarked, the integral (22) has a meaning even if the derivatives are only required to exist almost everywhere. In order to insure the convergence and finiteness of the Dirichlet integral, Beppo Levi introduced a class of admissible functions characterized by the following conditions:

(1) u is continuous in $\bar{\Omega}$.
(2) u has a well-defined Dini derivative (any one of them) with respect to x (y) on almost all lines $y = \text{const.}$ ($x = \text{const.}$).
(3a) $\int_{x_0}^{x_1} \partial u(x, y)/\partial x\, dx = u(x_1, y) - u(x_0, y)$, for a.a. lines $y = \text{const.}$
(3b) Analogously for x and y interchanged.
(4) u has the prescribed values on the boundary $\partial\Omega$.
(5) The function u has a finite Dirichlet integral (22).[34]

Levi showed how from a given minimal sequence in this space one could find a uniformly convergent minimal sequence, the limit of which would give the Dirichlet integral its minimal value. His primary motivation for the introduction of the general space of admissible functions was that it was easier to find the minimal function in the larger space. However this generalization did not introduce new solutions, for as Levi showed the minimizing function was in fact a C^2 function and thus a solution to the Dirichlet problem.[35] Beppo Levi's spaces were used by G. Fubini (1879–1943) [1907] in a very similar way.

Though Levi's admissible functions were obviously defined to suit the specific problem they were designed for, they have a close logical relationship to Tonelli's absolutely continuous functions. Indeed, Beppo Levi's admissible functions in a square are absolutely continuous in Tonelli's sense.[36] If we omit condition (4) Beppo Levi's space gives a definition of the Sobolev space H_2^1 (see note 29).

Tonelli did not explicitly state the relationship between his absolutely continuous functions (§20) and Beppo Levi's spaces, not even in his later application of them [1929] in variational calculus.[37]

25. But Levi's work was not forgotten. It was taken up by Otto Nikodym (born 1887) in 1932 in a talk, "Sur le principe du minimum", presented at the II Congrès des Mathématiciens Roumains in Turnue Sevrin and later published [1935].[38] Nikodym worked with a general class of differential equations including Laplace's equation. He used the structure of the spaces in a much more profound way than did Levi and Fubini. Where they had only operated with pointwise or uniform convergence he used the Dirichlet integral as a pseudonorm and thus obtained à very strong functional analytical tool for the solution of the Dirichlet problem and related problems. Further Nikodym studied the spaces "non seulement à cause de leur importance [in the solution of the Dirichlet principle] mais surtout parce qu'elles sont intéressantes en elles mêmes" [Nikodym 1933a, p. 129] and thus subjected them to a separate examination.

He removed Levi's boundary value condition (4) and his continuity condition (1) and defined (in three instead of two variables):

> A function is a "fonction de M. Beppo Levi" or shorter, a "fonction (BL)" in a box $D =]a_1, b_1[x]a_2, b_2[x]a_3, b_3[$ if

(1°) f is defined a.e. in D.
(2°) $f(\bar{x}, \bar{y}, z)$ is absolutely continuous in the z variable for a.a. fixed values $(\bar{x}, \bar{y}) \in]a_1, b_1[x]a_2, b_2[$ and similarly with the variables permuted.
(3°) The derivatives $\partial f/\partial x$, $\partial f/\partial y$, $\partial f/\partial z$ are in $L^2(D)$.[39]

Nikodym presented a series of theorems about the BL functions, among them that a BL function in D is in $L^2(D)$.[40] Most of the theorems were related to the topological behaviour of the BL space when equipped with the Dirichlet pseudonorm

$$\|f\|_D^2 = \iint\limits_D \left(\frac{\partial f}{\partial x}\right)^2 + \left(\frac{\partial f}{\partial y}\right)^2 + \left(\frac{\partial f}{\partial z}\right)^2 \qquad \text{(the } D \text{ norm).} \qquad (23)$$

Of course one cannot expect a D convergent sequence f_n to be convergent in any of the usual norms, but Nikodym showed that there exists a sequence of numbers a_n such that $f_n + a_n$ will converge in the L^2 norm on all closed subspaces of D.

Probably the most important result in his article states that a BL space is complete when equipped with the Dirichlet norm.[41,42]

In Nikodym's paper one encounters a new attitude toward the study of generalized derivatives. As already mentioned he considered the study of spaces of functions with generalized derivatives with certain properties as an interesting discipline in itself, independent of the applications, which had previously been its only *raison d'être*. These same tendencies were being exhibited during this period in the theory of partial differential equations as well.

26. As mentioned in §22 a totally different extension of the set of admissible functions in the calculus of variation was introduced by L. C. Young [1933] and developed further in [Young 1938]. By using the general Denjoy integral, Young generalized the differential operator to a subspace of the generalized absolutely continuous functions more extensive than the space of absolutely continuous functions. Moreover, Young's treatment led to a redefinition of the derivative of such functions so that many differential quotients were associated with one absolutely continuous function. However, let us look more closely at Young's ideas in order to understand what L. Schwartz meant when he wrote in his historical introduction [1950]:

> Par example les surfaces généralisées de L. C. Young utilisées en calcul des variations... procèdent d'idées analogues [to the theory of distributions].

Young's intuitive idea was that a derivative y' at a point x should not be given one fixed value but should be described as a distribution of likelihood, representing the probability with which $y'(x)$ takes on values in a finite [1933] or infinite [1938] set. In the [1933] article he compared this idea with similar ideas in quantum mechanics. In the continuous case this idea was formulated precisely in the following way [1938, §8]:

> By a *generalized admissible curve* we shall mean two finite functions measurable (B), $y(x)_*$, $y'(x, \alpha)_*$ the latter defined for $0 \le \alpha \le 1$ as well as for $x_0 \le x \le x_1$ such that
>
> $$y(x)_* = y_0 + \int_{x_0}^{x} dt \int_0^1 d\alpha\, y'(t, \alpha)_* = y_1 - \int_{x}^{x_1} dt \int_0^1 d\alpha\, y'(t, \alpha)_*. \tag{24}$$
>
> and for which the integral
>
> $$\int_{x_0}^{x_1} dx \int_0^1 d\alpha\, f(x, y(x), y'(x, \alpha)_*) \tag{25}$$
>
> exists, each of these integrals being interpreted as a repeated integral in the general Denjoy sense.

The last condition was derived from the variational integral

$$\int_{x_0}^{x_1} f(x, y(x), y'(x))\, dx, \tag{21}$$

which was to be minimized, perhaps with certain boundary value conditions. One can think of $y'(x, \alpha)_*$ as a kind of generalized derivative of y_* (Young did not use these words). Equation (24) shows that $y(x)$ is approximately differentiable with the approximate derivative $y'(x)_*$ given by

$$y'(t)_* = \int_0^1 d\alpha \, y'(t, \alpha)_* \quad \text{a.e.} \tag{26}$$

Evidently, the connection between this and the intuitive consideration is that for fixed t the probability that the derivative lies between u and v is given by $y_*'^{-1}[u, v]$; the mean value

$$\int_0^1 d\alpha \, y'(t, \alpha)_*$$

of $y'(t, \alpha)_*$ gives the ordinary derivative according to (24).

Young showed that in this way he could minimize variational integrals for which the greatest lower bound was not attained in the classical theory.

> For instance, in the simple example, due to Caratheodory [1906]
>
> $$f(x, y, y') = \frac{(1 + y^2)(1 + y'^2)}{\{2(1 + y'^2)\}^{1/2} - 1}, \qquad x_0 = y_0 = y_1 = 0, x_1 = 1. \tag{27}$$
>
> In this example we have $f \geq 2$ with equality only if $y = 0$ and $y' = \pm 1$. Clearly, for an ordinary curve $y(x), y'(x) \ldots$ the conditions $y(x) = 0, y'(x) = \pm 1$ at almost all points x are not compatible. [Young 1938, p. 249.]

The integral (21) obviously has 2 as its greatest lower bound, and according to Young's argument, this is not attained for ordinary curves. Young continued:

> The minimum of the generalized problem [i.e. the minimalization of (25) with an admissible curve] clearly has the same value, but this value is attained when we choose for all x,
>
> $$y(x)_* = 0, \qquad y'(x, \alpha)_* = \begin{cases} -1(\alpha < \frac{1}{2}), \\ 1(\alpha \geq \frac{1}{2}). \end{cases} \tag{28}$$

27. Young's theory is thus radically different from the other generalizations treated in this chapter in that its aim is not to extend the differential operators. In this respect it differs from the theory of distributions as well. But in another way Young's theory is much closer to the theory of distributions than any of the other generalizations of this chapter. That is, Young extended the class of curves just as the function concept is extended in the theory of distributions. E. J. McShane, who in [1933] and [1940] gave an alternative definition of Young's generalized curves, expressed this idea very clearly [McShane 1940]:

> Thus we see that our generalized curves form an extension of the class of ordinary curves in the same way that for example the real numbers form an extension of the class of rationals.

These are the "analogous ideas" which L. Schwartz saw in Young's work. In the theory of differential equations the corresponding step was taken in 1936 (i.e. between Young's two papers) by Sobolev.

Part 5. Generalized Solutions to Differential Equations. Potential Theory

28. During the first period of the history of generalized solutions to partial differential equations, delineated in §3, hyperbolic equations mainly entered the discussion. However, when the problems of generalized solutions were taken up again, after the long period of inactivity during most of the nineteenth century,[43] this happened in connection with elliptic equations; mainly those most important in physics, the Laplace and the Poisson equations.[43a]

The Swede Henrik Petrini gave a clear explanation of the problem which motivated the introductions of the generalized solutions [1908]:

L'équation de POISSON

$$\Delta V = -4\pi\rho, \tag{29}$$

où V désigne le potentiel Newtonien dans le point $P(x, y, z)$ d'une masse à trois dimensions dont la densité en ce même point est égale à ρ,[44] a été déduite par POISSON en 1813 sous la condition que la densité est constante dans le voisinage du point P. Puis GAUSS, en 1840, a déduit la même formule sous la conditions plus générale, que ρ admette les dérivées du premier ordre. Après GAUSS plusieurs géomètres et physiciens, parmi lesquelles nous citons DIRICHLET, RIEMANN, CLAUSIUS, KIRCHHOFF, KRONECKER, ont essayé de déduire la formule de POISSON dans des cas plus généraux. Mais ce n'est qu'en 1882 que M. HÖLDER a réussi à la déduire sans avoir recours à la condition de GAUSS, en supposant seulement la condition

$$|\rho - \rho_0| < Ar^{\mu}, \tag{30}$$

ρ_0 étant la densité au point considéré P, ρ la densité dans un point quelconque Q, situé dans l'intérieur d'un petit espace autour du premier point, r la distance PQ, A et μ des constantes positives. Plus tard, en 1887 M MORERA, a déduit la même formule sous la condition plus générale encore, que l'intégrale

$$\int_0^r (\rho - \rho_0)\frac{dr}{r} \tag{31}$$

est finie et déterminée le long de chaque rayon vecteur partant du point P, et que ρ_0 est une constante absolue. Enfin j'ai réussi en [1899] à déduire la formule de POISSON sous la seule condition que la densité ρ est continue, mais en définissant le symbole ΔV de la manière suivante:

$$\Delta V = \lim_{\substack{h_1=0\\h_2=0\\h_3=0}} \sum \frac{1}{h_1}\left[\frac{\partial V(x+h_1, y, z)}{\partial x} - \frac{\partial V(x, y, z)}{\partial x}\right], \tag{32}$$

en *supposant que les rapports des incréments* h_1, h_2, h_3 ne tendent ni vers zéro ni vers l'infini. Dans ce cas il peut arriver que ΔV existe quoique les dérivées $\partial^2 V/\partial x^2$, $\partial^2 V/\partial y^2$, $\partial^2 V/\partial z^2$ n'existent pas séparément.

The above generalization of the Laplace operator and hence of the Poisson equation is, as far as I know, the first explicit generalization of a differential equation following the unsuccessful attempts in the eighteenth century. It is similar to the Riemann and Lagrange generalizations of the second-order derivative (14) and (17) in that it only requires the existence of one limit instead of two or three.

Petrini [1908] continues:

> Cependant tous ces développements ont été faits en supposant que la densité ρ est continue. Ils ne suffisent donc pas pour mettre en évidence, comment se comportent les dérivées secondes dans un point de la surface du corps. Pour cela il faut faire l'analyse des dérivées du potentiel dans le cas plus général encore, où la densité ρ peut être discontinue.

In the general case, in which ρ is only assumed finite and integrable, Petrini arrived at the nice result that the Newtonian potential satisfies the equation

$$\Delta V = -4\pi\rho + \theta, \tag{33}$$

where θ is a Riemann null function[45] (Δ being defined by (32)).

29. Another method, which in the long run proved to be more fruitful, was introduced by the Harvard Professor Maxime Bôcher and later followed up by other American mathematicians. The method, which is based on Green's theorem

$$\int_\Omega (u\,\Delta v - v\,\Delta u)\,d\bar{x} = \int_{\partial\Omega}\left(v\frac{\partial u}{\partial n} - u\frac{\partial v}{\partial n}\right)ds \tag{34}$$

was explained very clearly in Bôcher's paper, "On harmonic functions in two variables" [1905/06]:

> A function $u(x, y)$ is said to be harmonic throughout a certain region T of the x, y-plane if (1) it is analytic there, and (2) it satisfies Laplace's equation. It is well known that the first of these requirements may be replaced by the apparently much less restrictive one that the function u be continuous and have continuous first and second partial derivatives. In fact, it is sufficient to demand even less than the continuity of the second partial derivatives; but some demand beyond the mere existence of these derivatives must be made in order that it should be possible to apply Green's theorem.
>
> It is my object in the present note to show how the theory of harmonic functions in two dimensions may be established without demanding even the existence of the second partial derivatives. This makes it necessary for us to take as our point of departure not Laplace's equation, since this involves the second partial derivatives whose existence we do not wish to demand, but some other characteristic property of harmonic functions. We select for this purpose the fundamental property that the integral of the normal derivative of a harmonic function, extended around a closed curve, is zero, and we thus have as our starting point not a differential equation of the second order, but a differentio-integral equation which involves only first partial

derivatives. It is needless to insist on the fact that it is this differentio-integral equation and not Laplace's equation which forms the starting point in most physical applications of harmonic functions.

We will lay down the following:

Definition. *$u(x, y)$ is said to be harmonic throughout a region T of the x, y-plane if it is single valued and continuous there, has continuous first partial derivatives, and, for any circle which lies wholly within T, satisfies the equation*

$$\int \frac{\partial u}{\partial n}\, ds = 0 \tag{35}$$

the integral being extended around the circle, n denoting the exterior normal and s the arc.

It is clear that this includes all functions which are harmonic according to the ordinary definition. It is not so clear that it does not also include other functions; for since we have not demanded even the existence of the second derivatives of u, we cannot pass from (35) to Laplace's equation by the familiar applications of Green's theorem.

The aim of the paper was to show that the new definition did not include other functions than those which are harmonic according to the ordinary definition. The main theorem in the paper thus stated:

If $u(x, y)$ is harmonic within any region T, it is analytic and satisfies Laplace's equation at every point within T.

This theorem is a special case of the general theorem that Schwartz proposed: every distribution which satisfies Laplace's equation is an analytic function. Therefore one will not find new solutions to Laplace's equation when it is generalized. Such a general insight was already gained by Evans [1927] who expressed it in the following way:

... any equation which seeks to express in more general language the physical idea behind Laplace's equation will imply Laplace's equation itself.

Thus the motive behind Bôcher's generalization was not to find more general solutions. What then did he seek? It appears from the quotation that he was led by aesthetic motives to seek generality merely for the sake of mathematical beauty. And in fact he succeeded in proving the above theorem, which is more general and more striking than the classical result.

Bôcher's generalization was discovered independently by the German mathematician Koebe in a paper [1906] of the same year as Bôcher's paper was published. Koebe went even further than Bôcher proving that the following three conditions on a real function $u(x, y)$ or $u(x, y, z)$ are equivalent.

(1) $u \in C^2$ and $\Delta u = 0$.

(2) $u \in C^1$ and $\int_S (\partial u/\partial n)\, ds = 0$ for all sufficiently regular curves (surfaces) S.

(3) $u \in C^0$ and satisfies Gauss' mean value property:

For each point P and each ball B with center at P the mean value of u over the surface S of B is equal to $u(P)$, i.e. $u(P) = [1/A(S)] \int_S u\, ds$, where $A(S)$ is the area of S.

Despite the greater generality of Koebe's result, both with respect to number of dimensions and differentiability conditions his work, in contrast to Bôcher's, does not seem to have been followed up. There are pure mathematical reasons for the lack of further development of the most general condition (3) since it is not as easily applicable to other partial differential equations as condition (2).

30. Bôcher's generalization of Laplace's equation is based on a simple idea which we will find in all later generalizations. The idea is, for a given nth order (partial) differential equation A to find another equation or condition B which for C^n functions is equivalent to the original equation, but which makes sense for less regular functions as well. Functions which satisfy B can then be called generalized solutions to A.[46] As a rule the derivation of the new equation uses some form of partial integration, for example, the integral formulas of vector analysis: Green's, Gauss' or Stokes' formulas.

In this way the derived equation will usually be of the form: the equation B holds for all objects of a certain kind. These objects can be called test objects. In Bôcher's case the test objects were circles. We will later meet other kinds of *test curves*. Another set of test objects are the *test functions* which are at the heart of the theory of distributions.

31. Bôcher's ideas were continued by his fellow-countryman, G. C. Evans (born 1887), professor at Rice Institute. Evans worked a great deal in potential theory in which he was the first to use Lebesgue–Stieltjes integrals for studying potentials of arbitrary mass—or charge distributions in the plane and in space. In his first published account of his theories [1920] Evans gave the following sketch of the source of his ideas:

> These studies originated in 1907, when it first became apparent to me that the [potential] theory was unnecessarily complicated by the form of the Laplacian operator, but I did not work on the subject until 1913 when it occurred to me to use instead of the operator
>
> $$\left(\frac{\partial^2}{\partial x^2} + \frac{\partial^2}{\partial y^2}\right)u, \tag{36}$$
>
> the operator
>
> $$\lim_{h=0} \frac{u(x+h, y) + u(x, y+h) + u(x-h, y) + u(x, y-h) - 4u(x, y)}{h^2} \tag{38}$$
>
> or the operator
>
> $$\lim_{\sigma=0} \int_S \frac{\partial u}{\partial n}\, ds, \tag{39}$$
>
> where S is a closed contour containing an area σ which is allowed to approach zero.

This quote requires some comment. In the paper [1920] Evans dealt with potential theory in the plane "but much of the material is obviously independent of the number of dimensions". Therefore his formulas differed from those of his predecessor. But there are more differences than those which stem from the difference in dimension. A comparison of formula (38) and Petrini's definition (32), to which Evans referred, shows that he had taken one more step in the direction of generalization since his formula did not even presuppose the existence of the first-order derivative. Evans' definition (38) is a combination of Petrini's definition (32), which combined the x and y limits into one, and Riemann's procedure (17) in which the first and second derivative were treated together. Formula (39) is obviously closely related to Bôcher's formula (35), but contrary to Bôcher, Evans generalized the Laplace operator and not only Laplace's equation.[47] In his other papers, however, he generalized differential equations as did Bôcher. He indicated that he discovered the ideas independently of his predecessors.

Evans decided to use the definition (39): "of the two it is obviously the concept which is the more closely allied with the physical interpretation" [Evans 1920].

32. Even though Evans developed his ideas on generalized solutions to partial differential equations in connection with potential theory he published them for the first time in connection with the parabolic operator

$$\frac{\partial u}{\partial t} - \frac{\partial^2 u}{\partial x^2}. \tag{40}$$

This occurred in the article "On the reduction of integro-differential equations" of [1914]. In it he also mentioned the possibility of generalizing (40) on the same principles as Petrini but continued:

> It is more convenient for us, however, because closer to a possible physical interpretation and less laborious in analysis to define the operator by substituting for the parabolic differential equation
>
> $$\frac{\partial u}{\partial t} - \frac{\partial^2 u}{\partial x^2} = f(x, t) \tag{41}$$
>
> the generalized equation
>
> $$\int_S \left[\frac{\partial u}{\partial x} dt + u\, dx\right] = \iint_\sigma f(x, t)\, dx\, dt. \tag{42}$$
>
> in which S is an arbitrary closed curve of a kind, later to be described and σ the inclosed region.

Evans' test curves, or standard curves as he called them, were not necessarily circles as were Bôcher's but they constituted a larger class of curves Γ, the exact definition of which is not of concern here (defined in Evans [1914, p. 486; 1920, p. 261]).

In this way Evans had generalized the differential equation (41) to another equation (42) which made sense for all functions u which, together with their first derivative $\partial u/\partial x$, were continuous in and on the boundary of the considered domain Ω. He called such functions regular. The generalization allowed him to solve the following initial boundary value problem: Given a domain Ω as shown in Figure 5 and given on the darkened portion of the boundary a function α which is only assumed to be continuous with continuous derivatives. Moreover, assume that $f(x, t)$ is continuous and finite in Ω. Then (42) has a solution with α as its boundary value.

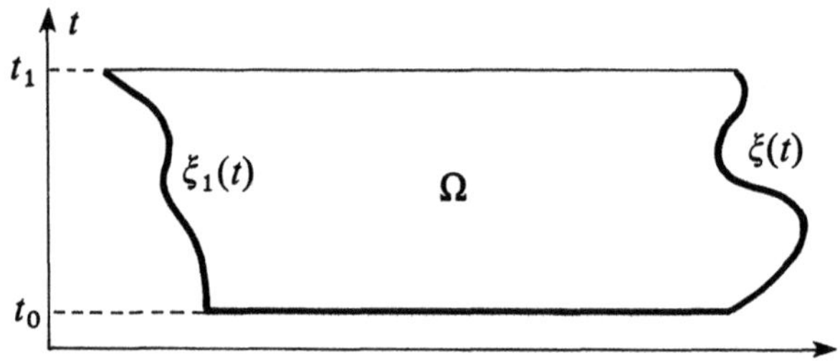

Figure 5

Hence Evans' [1914] paper gave the first rigorous proof that the solution of an initial—or boundary value problem could be generalized by generalizing the equation and relaxing the differentiability conditions. In so doing he indicated that Euler's point of view in the discussion of the vibrating string could be justified rigorously. However, Evans chose to use his generalization procedure on a parabolic differential equation instead of on the hyperbolic wave equation; he did not even mention the vibrating string, a fact which indicates that he did not consider the wave equation an especially significant equation. Among other mathematicians the eighteenth-century debate was still alive and the vibrating string had continued to have a magic sound (see Schwartz' remark §2 and Wiener's §43). It was still the standard example of an initial value problem.

33. After this digression on parabolic equations I shall now return to potential theory which was the starting point for Evans' work. In [1920] he used Green's theorem to transform Poisson's equation:

$$\Delta u(x, y) = F(x, y) \tag{43}$$

$$\int_S \nabla_n u \, ds = \int_\sigma F(x, y) \, dx \, dy \quad \text{for all } \sigma \text{ with } S = \partial\sigma \in \Gamma \tag{44}$$

corresponding to Bôcher's generalization of Laplace's equation ($\nabla_n = \partial/\partial n$ is the normal component of the gradient vector, and Γ is the same class of standard curves as in the 1914 paper). But Evans wanted to generalize (44) further and to eliminate the condition that u be once differentiable. To accomplish this he gave two methods for the generalization of the first-order derivatives.

The first method generalized the gradient vector directly in $\mathbb{R}^2$:

Let $\phi(M)$ be a vector point function whose component ϕ_α in every fixed direction α is summable superficially and $u(M)$ be a scalar point function summable superficially and such that $\int u(M)\, d\alpha$ may be defined for every given direction α. Then $\phi(M)$ is spoken of as the gradient of $u(M)$ and $u(M)$ as a potential function of $\phi(M)$, provided the equation

$$\int_\sigma \phi_\alpha d\tau = \int_S u\, d\alpha' \tag{45}$$

is satisfied for every S of Γ and for every fixed direction α, the direction α' being fixed and in advance of α by $\pi/2$.

For C^1 functions (45) is the well-known Green's formula for the plane.

The second method generalized the partial derivatives as follows:

We say that $D_\alpha u$, the generalized derivative in the direction α of $u(M)$ is the limit if such limit exists of the expression

$$D_\alpha u = \lim_{\sigma \to 0} \frac{1}{\sigma} \int_S u\, d\alpha; \tag{46}$$

where the fixed direction α' makes an angle $\pi/2$ with the fixed direction α and σ denotes the area enclosed by S; it is understood that σ tends to 0 in such a way that the ratio σ/d^2 where d is the diameter of σ, remains different from 0 by some positive quantity.

Evans found the relationship between the two definitions by using the theory of point set functions. He proved that if the gradient $\phi(M)$ exists then there exists a null set such that

$$D_\alpha u(M) = \phi_\alpha(M)$$

for all α and for all points M outside the null set. In this case Evans called u "the potential of its generalized derivative".

This new extension of the gradient operator made it possible for Evans to treat mass distributions which were more singular than those considered by Petrini, namely those distributions which were given by an additive set function (a measure) $f(e)$. Evans could prove that the logarithmic potential

$$u(M) = \frac{1}{2\pi} \int_\Sigma \log \frac{1}{MM'}\, df(e') \tag{47}$$

(which is the two-dimensional analogue of the Newtonian potential) satisfied

$$\int_S \nabla_n u\, ds = F(S) \qquad \forall S \in \Gamma \qquad \text{(cf. (44))} \tag{48}$$

where $\nabla_n u$ is the normal component of the *generalized* gradient vector and where $F(S)$ equals the value of f on the surface bounded by S plus a suitable mean value of f on S.[48] Evans' theorem could deal with point masses and mass distributions on curves and thus represented a significant generalization of the theorems proved by Petrini and Weyl, which I shall return to in §37.[49]

34. In [1920] Evans analyzed the relationships between his generalized derivatives and differentiation a.e. Among other things he showed that if u is the potential of its generalized derivative in Ω then there will exist a point (x_0, y_0) in Ω such that the function

$$\bar{u}(x, y) = u(x_0, y_0) + \int_{y_0}^{y} D_y u(x_0, \eta)\, d\eta + \int_{x_0}^{x} D_x u(\xi, y)\, d\xi \qquad (49)$$

is equal to u a.e. and $\partial\bar{u}/\partial x$ exists and is equal to $D_x u$ a.e. (do. for $\partial\bar{u}/\partial y$).

After Tonelli had defined his absolutely continuous functions in 1926 (see §20), Evans [1933, Appendix I] discussed how these were related to his own earlier generalizations. By a simple argument which used only Fubini's theorem and the identity (49) proved earlier he showed that:

> if in the definition of potential function of generalized derivatives the function is assumed to be continuous, as a point function, the specialized concept thus obtained is identical with the one formulated by Tonelli.... Thus it happens that the theorems proved by Tonelli in the last cited reference [Tonelli 1928/29] are in their essence special cases of those given earlier by the author.[50]

Evans' comparison is very interesting since it drew the first connection between two definitions of generalized derivatives which had arisen from completely different sources: measure and integral theory (more specifically the measure of surfaces) and potential theory. Even the very close relationship between the Beppo Levi spaces developed in the calculus of variations and the absolutely continuous functions had not been explicitly mentioned before this. Although Evans' comparison marked the beginning of the unification of the different trends in the theory of generalized derivatives and generalized solutions to differential equations, many *ad hoc* definitions continued to be given in connection with various problems without any attempt to compare them with other definitions. Generalized differentiation was not yet a theory.

35. Evans' methods were extended to n-dimensional spaces, and the theory was further developed in 1940 by two other Americans, J. W. Calkin and C. B. Morrey Jr., in their joint article "Functions of several variables and absolute continuity, I and II", (Calkin wrote part I [1940a] and Morrey part II [1940]). The two authors did to Evans' theory what Nikodym had done to Beppo Levi's; they separated it from potential theory and extended it to form a separate whole. But they did not conceal their motives for cultivating this area:

> While the results which are described here are not without intrinsic interest their chief importance for us lies in the uses to which they are put in researches to be described in subsequent papers: the author of the present paper [Calkin] has found them necessary for the study of partial differential equations by means of the theory of transformations in Hilbert space and the author of the following paper... [Morrey] has found them necessary for certain investigations in the calculus of variations.

Thus the applications of generalized differentiation in the two disciplines—the theory of differential equations and the calculus of variations—were united.[51] Instead of Evans' term "potential functions of their generalized derivatives":

> we shall call these functions merely functions of class $\mathfrak{P}$, sacrificing in the interests of brevity the descriptiveness of Evans' terminology.

As can be seen in the title of the paper the authors used Tonelli's term "absolute continuity" as well.

Calkin [1940a] called a function f defined in an open set G essentially absolutely continuous (E.A.C.) in x_k if

(1) $f(x_1, x_2, \ldots, x_n)$ is measurable in G and integrable in all closed boxes $\prod_{i=1}^{n} [a_i, b_i]$ included in G.

(2) There exists a function $g_k(x)$ (the generalized derivative) satisfying condition (1) such that

$$\int f(x_1, x_2, \ldots, x_{k-1}, b_k, x_{k+1}, \ldots) - f(x_1, \ldots, x_{k-1}, a_k, x_{k+1}, \ldots)\, dx'$$

$$= \int_{\prod [a_i, b_i]} g_k(x)\, dx \tag{50}$$

for almost all boxes $\prod [a_i, b_i]$ (as in note 49) where the first integral is to be taken over the product of the $[a_i, b_i]$s except for the kth.

A function which is (E.A.C.) in all its variables was said by Calkin to be of class $\mathfrak{P}$. $\mathfrak{P}$ constitutes a natural generalization of Evans' functions which are potential functions of their generalized derivatives, as this term was defined in Evans [1928] (see note 49). Calkin showed that $\mathfrak{P}$ functions can be approximated with Lipschitz' continuous functions in the following sense:

> A necessary and sufficient condition that a function $f(x)$ be of class $\mathfrak{P}$ on G is that there exist functions $g_1, \ldots, g_n$ measurable on G and absolutely integrable on every closed cell $[a, b]$ $[=\prod [a_i, b_i]]$ interior to G and for each such cell a sequence $\{f_q\}$ each member satisfying a uniform Lipschitz condition on $[a, b]$ such that

$$\lim_{q \to \infty} \int_{[a,b]} \left(|f - f_q| + \sum_{k=1}^{n} \left| g_k - \frac{\partial f_q}{\partial x_k} \right| \right) dx = 0. \tag{51}$$

Thus, conversely, if a sequence of Lipschitz continuous functions converges in the manner described above, then the limit function will be a $\mathfrak{P}$ function with the generalized derivative equal to the limit of the derivatives.[52] It was this theorem, wrote Calkin, which gave the $\mathfrak{P}$ spaces their great importance. We shall see how several mathematicians both before and after 1940 proved similar theorems which established the equivalence between generalized derivatives defined in terms of sequences and generalized derivatives defined in other ways.

36. Naturally it is tempting to consider the integral in (51) taken over G to represent a norm, but since integrability is only assumed in closed subboxes of G this is not possible in $\mathfrak{P}$. Calkin circumvented this difficulty by introducing a new class of spaces $\mathfrak{P}_\alpha$; at the same time he extended the theory from a L^1 to a L^α theory. The class $\mathfrak{P}_\alpha$ was defined to consist of all functions f in $\mathfrak{P}$ for which the expression

$$\overline{D_\alpha}(f) = \int_G \left(\sum_1^n |D_{x_k} f|^2\right)^{(1/2)\alpha} dx + \int_G |f|^\alpha \, dx \tag{52}$$

was finite. D_{x_k} denotes the generalized derivative. The main theorem in Calkin's part of the paper then asserted that $\mathfrak{P}_\alpha$, equipped with the norm $\overline{D_\alpha}$, was a Banach space and $\mathfrak{P}_2$ was a Hilbert space. Moreover, he introduced a series of other spaces and proved the relationship between these and the above-defined spaces.

Morrey [1940] continued the investigation in the second part of the paper and gave many theorems concerning change of variables [§6], boundary values [§7 and §9] and weak convergence [§8]. It is impossible even to indicate the variety of theorems which Morrey proved.

On the other hand, it is interesting to note a difference in the approaches of Evans and Calkin–Morrey: where Evans was mainly interested in the separate functions, Calkin and Morrey were more interested in the structure of different function spaces. This difference was very much in keeping with the development of analysis as a whole (see Ch. 1). Thus, though historically Calkin–Morrey's paper fits very nicely into this chapter, mathematically it is more strongly linked with the Hilbert space theory, to which I shall return in §51. However, in their 1940 paper Calkin and Morrey were apparently not influenced by other mathematicians working on the theory of differential operators in Hilbert spaces.[53]

37. After this series of Americans we shall turn to a European contribution to the use of generalized derivatives in potential theory. In a footnote to the paper "Über die Randwertaufgabe der Strahlungstheorie" of [1913] Hermann Weyl (1885–1955), after referring to Green's theorem (34), commented:

> Die für die mathematische Physik wesentliche Definition von Δv liegt nicht in der Gleichung
>
> $$\Delta v = \frac{\partial^2 v}{\partial x^2} + \frac{\partial^2 v}{\partial y^2} + \frac{\partial^2 v}{\partial z^2} \tag{53}$$
>
> sondern: für ein skalares oder vektorielles, stetig differentierbares Feld v ist Δv diejenige stetige Funktion (falls sie existiert) welche für jedes Raumstück J die Gleichung
>
> $$\int_J \Delta v \, dp = -\int_D \frac{\partial v}{\partial n} \, d\sigma \tag{54}$$
>
> erfüllt. [Weyl 1913, p. 182, footnote.]

Here $D = \partial J$.

This generalization of the Laplace operator immediately allowed Weyl to treat Poisson's equation:

Bei dieser Erklärung gilt beispielsweise für das Newtonsche Potential

$$v(p) = \int \frac{1}{r(pp')} f(p')\, dp' \qquad (r = \text{Entfernung}) \tag{55}$$

wie man sofort sieht, stets die Poissonsche Gleichung $\Delta v = -4\pi f$, wenn nur f stetig ist, während bei Zugrundelegung der gewöhnlichen Definition von Δv hierzu bekanntlich weitere Voraussetzungen über die Dichtigkeitsfunktion f nötig ist. [Weyl 1913, p. 182, footnote.]

Thus Weyl applied a test surface method (the three-dimensional analogue to the test curve method which Bôcher had already used for the Laplace equation in 1906) to obtain a result which had been proved by Petrini with a different method in 1899. More specifically Weyl's generalization is actually the same as Evans' [1920] generalized Poisson equation (44) with the classical meaning of the first derivative Δ_n. However, there is no indication of interaction between the ideas developed in parallel on both sides of the Atlantic. Nor does Weyl refer to Koebe.

38. From a footnote in a much later paper, "The method of orthogonal projection in potential theory" [1940], it appears that Weyl touched upon generalized derivatives when he lectured on vector analysis at the Technische Hochschule in Zürich during his appointment from 1913 to 1929. In the 1940 paper he included a special section dealing with these generalizations and, according to his own footnote, he "followed closely" the Zürich lectures.

In the [1940] article Weyl had set himself the task of solving the Dirichlet problem in the extended formulation S. Zaremba and Nikodym had given it (see note 38). Just as Nikodym did, Weyl solved the problem by operating with $f = \operatorname{grad} \varphi$ in L^2 (L^2 here describes square integrable vector functions $\|f\|_2 = \int |f_1|^2 + |f_2|^2 + |f_3|^2$) in which connection he noted:

Then the question arises how to characterize a vector field f as a gradient field without assuming more than its Lebesgue integrability. The vanishing of the line integral

$$\int (f\, dx) = \int (f_1\, dx_1 + f_2\, dx_2 + f_3\, dx_3) \tag{56}$$

over any closed curve in G will not do, because we have nothing but spatial integration at our disposal. The customary condition

$$\operatorname{rot} f = 0 \tag{57}$$

uses differentiation. Let v be any vectorial vanishing at the boundary of G.[54] The formula

$$\operatorname{div}[f, v] = v \operatorname{rot} f - f \operatorname{rot} v \tag{58}$$

for the vector product $[f, v]$ with its integrated consequence

$$\int (v \cdot \operatorname{rot} f) = \int (f \cdot \operatorname{rot} v) \tag{59}$$

shows that (57) is equivalent to the relation

$$\int (f \cdot \operatorname{rot} v) = 0 \tag{60}$$

holding for all fields of the above-described nature. This characterization satisfies our demands. [Weyl 1940.]

Weyl called such functions irrotational.

Similar to this generalization of (57) to (60) he generalized the equation

$$\operatorname{div} f = 0 \tag{61}$$

to

$$\int f \operatorname{grad} \psi = 0 \quad \text{for all scalar fields } \psi \text{ of class } \Gamma = C_c^1. \tag{62}$$

Functions satisfying (62) he called solenoidal.

Weyl introduced the subspace $\mathfrak{F}$ of $L^2(G)$ consisting of all irrotational vector fields, and the subspace $\mathfrak{E}$ of both solenoidal and irrotational vectors. The closure in $L^2(G)$ of fields of the form grad ψ where $\psi \in C_c^1(G)$ he called $\mathfrak{G}$. His generalization of Zaremba–Nikodym's theorem then assumes the form

$$\mathfrak{F} = \mathfrak{G} + \mathfrak{E}, \tag{63}$$

where the sum is an orthogonal sum in L^2.[55]

39. In the [1940] article Weyl also generalized differentiation along different lines:

Our next concern is the introduction of div and rot without the hypothesis of differentiability. We shall use the notations div*, rot* for these generalizations of the differential operators div, rot. The existence of

$$\operatorname{div}^* f = \rho \tag{64}$$

for a continuous field f means the existence of a continuous function ρ in G such that

$$\int_{T'} f \cdot n \, d\sigma = \int_T \rho \qquad (T' \text{ boundary of } T) \tag{65}$$

holds for any cube in a certain neighbourhood N of any point x^0 of G. . . . Source-free fields f are characterized by the condition

$$\operatorname{div}^* f = 0. \tag{66}$$

The existence of

$$\operatorname{rot}^* f = u \tag{67}$$

for a continuous field f means the existence of a continuous vector field u in G such that

$$\int_Q u \cdot n \, d\sigma = \int_{Q'} f \cdot dx \tag{68}$$

for any square Q in a certain neighbourhood N of any point x^0 in G.... Whirl-free fields f are characterized by the equation

$$\text{rot}^* f = 0. \tag{69}$$

Thus Weyl used Gauss' theorem to generalize div and Stokes' theorem to generalize rot. In this way he obtained two generalizations of each of the equations

$$\text{rot } f = 0 \quad \text{and} \quad \text{div } f = 0, \tag{70}$$

namely those described in §38 and those defined above. He proved that the definitions mentioned last implied those mentioned first, i.e. [Weyl 1940, p. 426].

$$\text{“Any source-free field } f \text{ is solenoidal} \ldots \tag{71}$$

$$\text{and any whirl-free field } f \text{ is irrotational.”} \tag{72}$$

The inverse theorems are only true for continuous fields [Weyl 1940, p. 426].

40. Weyl's two generalizations can be characterized as a test function method and a test surface method, respectively. The test surface method (§39) for defining rot* and div* is very similar to the method he used in 1914 to generalize the Laplace operator (§37), and it was probably the method Weyl used in his Zürich lectures. These probably also included generalizations to starred operators of well-known vector identities such as

$$\text{div}(\varphi f) = \varphi \text{ div } f + f \text{ grad } \varphi, \tag{73}$$

which Weyl proved in his 1940 paper [p. 425]. The test function definition (§38) on the other hand is of quite a different nature. If one can trust Weyl's words in [1940], this type of generalization had also been set forth in the Zürich lectures since the implications (71) and (72) are given in that section of the [1940] paper which according to Weyl closely followed his lectures. If that is true Weyl was the first to propose the test function generalization. For several reasons, however, I believe that Weyl did not use this procedure before his 1940 treatment of the Dirichlet problem.[56] If I am right Wiener can be credited with the invention of the test function method [1926b] (see §43).

41. I want to conclude this section on generalized solutions to elliptic equations by showing how two mathematicians came to realize the importance of some kind of generalized differentiation in the theory of subharmonic functions.

In T. Radó's booklet *Subharmonic Functions* [1937] which, according to M. Brelot [1972, p. 6], summarized the results obtained in the field, a subharmonic function in G was defined as a function u which is (1) upper semicontinuous and satisfies (2) for every subregion G' of G with $\bar{G}' \subset G$ and for all harmonic functions H for which $H \geq u$ on $\partial G'$ the inequality $H \geq u$ is valid in the whole of G'.

One of the main theorems in Radó's book concerned the "Representation of subharmonic functions in terms of potentials" [Ch. 6]. The theorem says that every subharmonic function u is a sum (locally)

$$u(P) = -\int_G \log \frac{1}{PQ}\, d\mu(e_Q) + H(P) \tag{74}$$

of a potential of a mass distribution $\mu(e_Q)$ in G and a harmonic function H.[57] Radó concluded the chapter:

> If u is a *smooth* subharmonic function then the corresponding distribution can be expressed in terms of the Laplacian Δu. It is then natural to expect that the preceding theorems can be discussed in terms of the generalized Laplacians introduced by various authors. It seems that no explicit expression was given as yet on this basis.

Radó referred to a note which F. Riesz had received from N. Wiener and which Riesz published [Wiener 1927]. This paper, the main aim of which was to prove a coordinate-independent version of the Riesz representation theorem, did not discuss the above theorem, but included other interesting considerations on subharmonic functions, among which was a more natural definition of this class of functions: Wiener remarked [1927, p. 8] that the operator

$$\frac{6}{h^2}\left\{\frac{1}{4\pi h^2}\iint\limits_{|r|=h} f(x+r)\,ds - f(x)\right\} \tag{75}$$

had the same relation to the Laplace operator Δ as

$$\frac{1}{h}(f(x+h) - f(x)) \tag{76}$$

had to d/dx. Hence he defined a subharmonic function as a function f for which (75) is nonnegative for all x and all $h > 0$. This is a test surface generalization of

$$\Delta f \geq 0, \tag{77}$$

the test surfaces being surfaces of spheres. He did not let h tend to zero to obtain a generalized Laplace operator, but even without such a step he was able to use the generalization (75) in a discussion of charge distributions concentrated in points and on curves and surfaces.

During the same year that Schwartz published his first monograph on the theory of distributions [1950/51], J. Deny showed in his paper, "Les potentielles d'énergie finie" [1950], how this new theory could be used in the theory of subharmonic functions, perfecting the ideas of Wiener.[58]

It is interesting to note that Schwartz' work on the distributions was a result of his interest in generalized solutions to elliptic equations (see Ch. 6, §3–5).

Part 6. Generalized Solutions to Hyperbolic Partial Differential Equations. The Cauchy Problem

42. Generalized solutions to hyperbolic partial differential equations were not rigorously defined before 1927, i.e. 28 years after Petrini had generalized Poisson's equation. This time lag may appear much more striking when one keeps in mind that the Cauchy problem for hyperbolic equations was the first problem which gave rise to generalization attempts and that this problem during the nineteenth century necessitated physical generalizations. An explanation for the astonishing time interval between the generalization of solutions in potential theory and in the theory of hyperbolic equations can be found in the development of the two disciplines themselves.

At the beginning of the twentieth century potential theory was no doubt far ahead of the rest of the theory of partial differential equations. An examination of the *Jahrbuch über die Fortschritte der Mathematik* reveals that in the period from 1916, when potential theory and elliptic equations became separated in the index from hyperbolic equations and their physical applications, until the Second World War, when this review magazine ceased to exist, the number of papers on potential theory far exceeded the number of papers on hyperbolic equations. When we add to this that potential theory forms a smaller and more well-defined field than the theory of hyperbolic equations it is obvious that potential theory was the more developed of the two. Therefore the essential questions concerning the regular solutions were more quickly given satisfactory answers in potential theory. For this reason mathematicians became interested in the irregular cases in this field earlier than they did in the hyperbolic case where many basic problems, for instance in connection with characteristics, fundamental solutions, etc., continued to occupy mathematicians for a longer time.

I do not want to give the impression that all problems relating to classical solutions were solved when generalized solutions were taken up, for this was certainly not the case. However, a mathematician's desire to generalize a theory undoubtedly becomes more intense when more of the great classical problems in the theory are solved.[59]

43. One of the methods used to solve initial value problems for hyperbolic partial differential equations, employed most often in practice in the beginning of the twentieth century, was Heaviside's operational calculus. In this calculus one operated very freely with differential operators, on both regular and irregular functions. The mathematicians found it difficult to explain why the method worked even in the regular cases [Lützen 1979]. Therefore, they mainly ignored the irregular cases for which reason questions of generalized solutions were usually not dealt with. (For the rather unsatisfactory treatment of the δ-function which arose in this connection, see Ch. 4, §21–27.)

Wiener's paper, "The operational calculus" [1926b], is an exception, since it treats generalized solutions to partial differential equations. In this paper

Wiener gave the operational calculus a rigorous basis by using generalized Fourier transformations (see [Lützen 1979 Ch. IV, 3] and Ch. 3, note 10). Wiener began the last section of the paper which is devoted to "The operational solution of partial differential equations" with the following observations:

> Before we enter on this topic in detail, it is important to consider the nature of the solution of a partial differential equation. Let us consider the linear equation
>
> $$A\frac{\partial^2 u}{\partial x^2} + B\frac{\partial^2 u}{\partial x\,\partial y} + C\frac{\partial^2 u}{\partial y^2} + D\frac{\partial u}{\partial x} + E\frac{\partial u}{\partial y} + Fu = 0, \tag{78}$$
>
> where for simplicity's sake, we shall suppose that the coefficients have as many derivatives as we shall need in the work which follows. If u satisfies this equation, it must manifestly possess the various derivatives indicated in the equation. As is familiar, however, in the case of the equation of the vibrating string, there are cases where u must be regarded as a solution of our differential equation in a general sense without possessing all the orders of derivatives indicated in the equation, and indeed without being differentiable at all. It is a matter of some interest, therefore, to render precise the manner in which a nondifferentiable function may satisfy in a generalized sense a differential equation.
>
> Let $G(x, y)$ be a function positive and infinitely differentiable within a certain bounded polygonal region R of the XY plane, vanishing with its derivatives of all orders on the periphery of R, and zero outside R. Then there is a function $G_1(x, y)$ such that
>
> $$\iint_R (Au_{xx} + Bu_{xy} + Cu_{yy} + Du_x + Eu_y + Fu)G(x, y)\,dx\,dy = \iint_R u(x, y)G_1(x, y)\,dx\,dy \tag{79}$$
>
> for all u with bounded summable derivatives of the first two orders, as we may show by integration by parts. Thus a necessary and sufficient condition for u to satisfy our differential equation almost everywhere is that
>
> $$\iint_R u(x, y)G_1(x, y)\,dx\,dy = 0 \tag{80}$$
>
> for every possible G (as the Gs form a complete set over any region), and that u possess the requisite derivatives. We can thus regard a function orthogonal to every G_1 as satisfying our differential equation in a generalized sense. This sense is more general than that developed by Bôcher[60] as not even the existence of first-order derivatives is postulated. If a sequence of generalized solutions of our differential equation converges in the mean over a given area to a function $\Phi(x, y)$, it is now manifest that Φ is itself a generalized solution of this differential equation over this area.

Wiener's mention of the problem of the vibrating string shows that in spite of the lack of mathematical attention for more than a hundred years the idea of generalized solutions to this equation had not been forgotten. However, he chose to exemplify his method by the telegraphers' equation

$$u_{tt} + (a + b)u_t + abu = u_{xx}. \tag{81}$$

His motivation for considering the generalized solutions to this equation and to the more general one (78) is apparent from his operational solution of (81) with the boundary condition

$$u(x = 0, t) = f(t) \tag{82}$$

for a sufficiently regular function f. Formally the operational solution to this problem is given by

$$u(x, t) = \exp[-x\sqrt{(d/dt + a)(d/dt + b)}]f(t). \tag{83}$$

Wiener's preceding treatment of the operational calculus, however, did not give a rigorous meaning to (83). On the other hand, he had shown that there existed a sequence $f_\lambda(t)$ converging to f in the mean for which

$$u_\lambda(x, t) = \exp[-x\sqrt{(d/dt + a)(d/dt + b)}]f_\lambda(t) \tag{84}$$

had a meaning, such that $u_\lambda(x, t)$ was a solution to (81) with the boundary value $u_\lambda(0, t) = f_\lambda(t)$. Moreover, Wiener showed that the u_λs converged in the mean to a function $u(x, t)$, which he then wrote as (83). After the limit of the u_λs was taken, Wiener could not be sure that the function u was a solution to the telegraphers' equation (81), but as he had remarked at the end of the quote u was a generalized solution to (81). Moreover, since the convergence of u_λ was uniform for x in any finite interval, he could conclude that u had the boundary value f. This shows why Wiener was forced to define generalized solutions.

44. Wiener's generalization was probably the first rigorous application of the test function method, the main idea of which is to transfer by partial integration (i.e. by adjoining the operator) the differential operator from the unknown function onto a test function (see, however, §38). One might expect that he would have given a sequence definition since it was precisely such a property that he needed in his operational calculus. Choosing another method made his definition appear less *ad hoc*; and he did remark that the test function definition implied the sequence definition.

However, when we compare this paper with Wiener's paper one year later [1927], which was discussed in §41, it becomes apparent that his treatment of generalized solutions to partial differential equations still possessed an *ad hoc* character. For in the 1927 paper he chose to use a type of test surface method without even mentioning the possibility of applying his year old method. These different approaches by the same mathematician show that problems about generalized solutions to partial differential equations were not yet seen as a part of one unified theory. The theories were still so closely linked with the problems they were set up to solve that their mutual connection was not easily seen.[61]

45. From the period preceding Sobolev's revolutionary papers [1935b, 1936a] I have found only one clear case in which a Cauchy problem alone motivated the generalization of the concept of a solution, similar to what had happened in a nonrigorous way in the discussion of the vibrating string. This Cauchy problem, which arose from hydrodynamics, was treated by Oseen [1911a] and Leray [1934], and the latter explained the problem in this way:

> La théorie de la viscosité conduit à admettre que les mouvements des liquides visqueux sont régis par les équations de Navier; il est nécessaire de justifier *a posteriori* cette hypothèse en établissant la théorème d'existence suivant: il existe une solution des équations de Navier qui correspond à un état de vitesse donné arbitrairement à l'instant initial. C'est ce qu'a cherché à démontrer M. Oseen.

The Swedish mathematician C. W. Oseen derived in 1911 three theorems which gave a partial solution to the above problem: He could prove that if a fluid at a given time is in a regular state, then the movement will continue to be regular in a certain time interval thereafter, and if the kinetic energy is finite in the initial state, then it continues to be finite in some time interval. In his first paper on the subject [1911a] the term "regular" referred to a movement in which all the derivatives in the Navier–Stokes equations were continuous. In a later paper in the same volume of *Arkiv för Matematik, Astronomi och Fysik*, he demonstrated the same theorems for a more extensive class of "regular functions".

> Nun lassen sich die Navierschen Differentialgleichungen in ein System von Integralgleichungen umsetzen. In diesen kommen die soeben aufgezählte Ableitungen [the derivatives in the Navier–Stokes equations] nicht vor, und man kann in der Tat leicht zeigen dass die Integralgleichungen Systeme von Lösungen besitzen, für welche diese Ableitungen überhaupt nicht existieren.

The above-mentioned equations are of the form:

$$u(x, y, z, t) = \int_{\Omega} F(u(\xi, \eta, \zeta, \tau))K(x, y, z, t, \xi, \eta, \zeta, \tau)\, d\xi\, d\eta\, d\zeta\, d\tau \tag{85}$$

plus some similar terms involving the first derivatives of u on the surface $\partial\Omega$.

This type of integral equation had not been proposed earlier as a way to generalize differential equations, and I have found no later application of the method either. Oseen continued:

> Man muss sich dann fragen ob diese Lösungen der Integralgleichung, welche nicht gleichzeitig Lösungen der Differentialgleichungen sind, nur eine mathematische Bedeutung besitzen oder ob einige unter ihnen auch eine physikalische Bedeutung haben. Ich stelle mir hier die Aufgabe zu zeigen dass letzteres der Fall ist und dass also nicht die Navierschen Differentialgleichungen sondern die Integralgleichungen als der adaequate Ausdruck der physikalischen Hypotesen zu betrachten sind, welche der Theorie der Bewegung reibenden Flüssigkeit zu Grunde liegen. (Cf. note 8.)

To accomplish this goal, Oseen derived his integral equations directly from the physical laws governing the fluid and not from the Navier–Stokes equations, as he, in the first part of the above quote, had indicated would be possible. In this respect Oseen's derivation was similar to the one used by Riemann, Christoffel and others (§14); the resulting equation (85), however, differed essentially from those obtained by the previous authors in that (85) determined the behaviour of the whole system; the equations found by Riemann and Christoffel only described the propagation of the singularity.

At the end of his paper [1911b] Oseen remarked that other parts of mathematical physics could be generalized and simplified in a similar way. As examples he mentioned potential theory and the theory of heat propagation, but he did not offer any details.

46. The French mathematician Jean Leray (born 1906) was not satisfied with Oseen's treatment of the "regular" solutions. Thus he wrote [Leray 1934][61a]:

> Il [Oseen] n'a réussi à établir l'existence d'une telle solution que pour une durée peut-être très brève succédant à l'instant initial...; mais il ne semble pas possible de déduire... que le mouvement lui-même reste régulier; j'ai même indiqué une raison qui me fait croire à l'existence de mouvements devenant irréguliers au bout d'un temps fini; je n'ai malheureusement pas réussi à forger un exemple d'une telle singularité.

Yet in order to prove that the Cauchy problem had a solution in some sense which would remain regular at all times, he generalized the Navier–Stokes equations in a way different from Oseen's. He proceeded along purely mathematical lines. He multiplied the Navier–Stokes equations

$$\nu\,\Delta u_i(x,t) - \frac{\partial u_i}{\partial t}(x,t) - \frac{1}{\rho}\frac{\partial p(x,t)}{\partial x_i} = \sum_{k=1}^{3} u_k(x,t)\frac{\partial u_i(x,t)}{\partial x_k} \tag{86}$$

(where u is the velocity and p is the pressure) by a sufficiently smooth, divergence-free vector, $a_i(x, t)$ and integrated over $[0, t]$ and $\mathbb{R}^3$. After partial integration this gave him [Leray 1934, p. 240]

$$\begin{aligned}\sum_i \iiint_{\mathbb{R}^3} u_i(x,t)a_i(x,t)\,dx &= \sum_i \iiint_{\mathbb{R}^3} u_i(x,0)a_i(x,0)\,dx \\ &\quad + \sum_i \int_0^t dt' \iiint_{\mathbb{R}^3} u_i(x,t')\left[\nu\,\Delta a_i(x,t') + \frac{\partial a_i(x,t')}{\partial t'}\right]dx \\ &\quad - \sum_i\sum_k \int_0^t dt' \iiint_{\mathbb{R}^3} u_k(x,t)u_{i,k}(x,t')a_i(x,t')\,dx,\end{aligned} \tag{87}$$

which was supposed to hold for all divergence-free vectors a_i (u_{ik} means $\partial u_i/\partial x_k$).

This gives a nice application of the test function method. Even more interesting, however, is what Leray writes after this: in order to avoid the assumption of the existence of $\partial u_i/\partial x_k$ in (87), he introduced what he called quasi-differential operators [Leray 1934, p. 205]:

Définitions des quasi-derivées: Soient deux fonctions de carrés sommable sur $\prod\, [=\mathbb{R}^3] U(y)$ et $U,_i(y)$; nous dirons que $U,_i(y)$ est la quasi-derivée de $U(y)$ par rapport a y_i quand la relation

$$\iiint \left[U(y)\frac{\partial a}{\partial y_i} + U,_i(y)a(y) \right] dy = 0 \tag{88}$$

sera vérifiée; rappelons que dans cette relation $a(y)$ représente une quelconque des fonctions admettant des dérivées premières continues qui sont, comme ces fonctions elles-mêmes de carrés sommable sur $\prod$.

This is again a test function definition very similar to Wiener's, and like Wiener, Leray was mainly interested in another property of the generalization, namely the following: any L^2 function U can be approximated by the sequence of C^∞ functions

$$U_n = U * \rho_n,$$

where ρ_n is a smooth approximating identity; if u has a quasi-derivative U_i then this is the weak L^2 limit of the derivatives $\partial U_n/\partial x_i$. This theorem and its converse[62] were very important to Leray because it enabled him to prove his main theorem on the existence of a generalized solution defined for all $t \geq 0$[63] by first solving some related but more regular problems and then passing to the limit.

Leray used the test function generalization in a third way, namely to generalize the divergence operator (see note 63). His consistent use of this generalization method occupies a central position in the prehistory of the theory of distribution as he taught it to his student L. Schwartz at the Ecole Normale.

47. The sequence property, which both Wiener and Leray found so valuable, was not always passed over as a definition. It was used by the American D. C. Lewis in his treatment [1933] of an initial-boundary value problem for the equation

$$\frac{\partial^2 u}{\partial t^2} - \frac{\partial^2 u}{\partial y^2} = F\left(\frac{\partial u}{\partial y}, \frac{\partial u}{\partial t}, u, y, t\right), \tag{89}$$

with $u(0, t) = u(\pi, t) = 0$, $u(y, 0) = f(y)$ and $u_t(y, 0) = g(y)$, where f has a derivative in L^2 and $g \in L^2$. Lewis compared this problem with the infinite system of ordinary differential equations which is found by formally inserting the Fourier series for u, f and g (expanded in the y variable) and their term-by-term differentiated series in (89), multiplying by $\sin n(\tau - t)$ $(n = 1, 2, \ldots)$ and integrating from 0 to π. In order to obtain complete equivalence between the two problems he had to generalize the equation (89). He defined a generalized solution to the second-order equation $P(u) = 0$ in

$$Q = \begin{cases} 0 \leq y \leq \pi, \\ 0 \leq t \leq K, \end{cases}$$

to be a function u for which $\partial u/\partial y$ and $\partial u/\partial t$ exist a.e. and for which there exists a sequence $u_k(y, t)$ in $C^2(Q)$ such that:

(1) $$\lim_{n\to\infty} u_n(y, t) = u(y, t) \quad \text{uniformly in } y \text{ and } t,$$

(2) $$\lim_{n\to\infty} \int_0^\pi \left[\frac{\partial u}{\partial y} - \frac{\partial u_n}{\partial y}\right]^2 dy = 0 \quad \text{uniformly in } t,$$

(3) $$\lim_{n\to\infty} \int_0^\pi \left[\frac{\partial u}{\partial t} - \frac{\partial u_n}{\partial t}\right]^2 dy = 0 \quad \text{uniformly in } t,$$

(4) $$\lim_{n\to\infty} \int_0^K \int_0^\pi [P(u_n(y, t))]^2 \, dy \, dt = 0.$$

The sequence of functions u_n which Lewis used in the proof of the equivalence of the two problems consisted of solutions to suitable finite subsystems of the infinite system of ordinary differential equations[64].

Lewis' generalization is a combination of the method using differentiation a.e. and the sequence definition. We will see that a pure sequence definition was used later by both Sobolev and Schwartz prior to their test function approach to the theory of distributions.

48. The sequence definition and the test function definition were both treated in Bochner's paper, "Linear partial differential equations with constant coefficients" [1946], the same year that Schwartz published his first paper on the theory of distributions.

Bochner chose to define a generalized solution to the differential equation

$$\Lambda f = 0, \tag{90}$$

where

$$\Lambda f = \sum_{r_1+\cdots+r_n\le N_0} a_{r_1\cdots r_n} \frac{\partial^{r_1+\cdots+r_n} f}{\partial x_1^{r_1}\cdots\partial x_n^{r_n}} \tag{91}$$

by the following sequence definition:

> We will say that $f(x)$ is a weak solution of class C^N in D if it is defined almost everywhere in D and Lebesgue integrable in every compact subset of D, and if corresponding to any point x^0 in D there exist a neighbourhood $U = U(x^0)$ such that in $U, f(x)$ is a weak limit of strict solutions of class C^N.[65] [Bochner 1946.]

He proved the following equivalence with the test function definition [Bochner 1946, theorem 2].

> If $f(x)$ is a weak solution of class C^N then
>
> $$\int_D f\Lambda\varphi \, dv_x = 0 \tag{92}$$
>
> for every testing function of class C^N.

In [theorem 7] he proved the converse.[66] Bochner used the term testing function for a function with compact support in D.[67] He proved several interesting theorems about the weak solutions, among others that a weak solution of Laplace's equation was an ordinary solution (after correction on a null-set) (This is Bôcher's 1906 theorem proved for the new type of generalized (weak) solutions); he also gave a sufficient condition under which a weak solution f in a subset D_0 of D could be completed to a weak solution $\bar{f}$ in all of D by setting $\bar{f} = 0$ in $D \setminus D_0$.

Bochner's theory for weak solutions was adopted in Bochner and Martin's book *Several Complex Variables* [1948] with small alterations. The most striking change was their use of the test function characterization as the definition.[68]

Bochner's [1946] article shows how far the theory of generalized solutions had developed before the theory of distributions had become known. Moreover, it illustrates the disconnectedness of the prehistory of distributions: Bochner, who had made the most outstanding contributions to the theory of generalized Fourier transformations (see Ch. 3, §12–18), apparently did not see the connection between that theory and the generalized solutions to partial differential equations.

49. The test function definition was the first definition of a generalized solution to a partial differential equation to appear in a textbook—in one of the most famous and influential textbooks of this century: Courant and Hilbert's *Methoden der Mathematischen Physik, II* [1937, pp. 469–470]. Here generalized solutions were introduced in order to show that discontinuities in the solutions to hyperbolic differential equations always progressed along characteristic surfaces.

Courant discussed a second-order homogeneous hyperbolic differential equation:

$$L[u] = \sum_{i,k=1}^{n} a_{ik} \frac{\partial^2 u}{\partial x_i \, \partial x_k} + \sum_{i=1}^{n} b_i \frac{\partial u}{\partial x_i} + cu = 0, \tag{93}$$

where a_{ik}, b_i, c need not be constants. The adjoint operator

$$M[v] = \sum_{ik} \frac{\partial^2}{\partial x_i \, \partial x_k} (a_{ik} v) - \sum_{i} \frac{\partial}{\partial x_i} (b_i v) + cv \tag{94}$$

had been treated already by Riemann (see, for example, Kline [1972, p. 692]) who had proved Green's theorem:

$$\int_G (vL[u] + uM[v]) \, d\bar{x}$$

$$= \int_{\partial G} \sum_i \left[v \sum_k a_{ik} u_k - u \sum_k \frac{\partial}{\partial x_k} (a_{ik} v + b_i uv \right] \frac{\partial x_i}{\partial n} \, d\sigma.^{69} \tag{95}$$

Wiener had implicitly used the adjoint operator and (95) to define his generalized solutions, but Courant was the first to use it explicitly. He remarked that if u and v were $C^2(G)$ functions for which v and its first-order derivatives vanished on the boundary, and u was a solution to (93) then

$$\int_G uM[v]\,d\bar{x} = 0. \tag{96}$$

Diese Relation gefordert für willkürliches v ist mit der Differentialgleichung $L[u] = 0$ äquivalent wenn u zweimal stetig differenzierbar ist. Sie behält aber ihren Sinn auch noch, wenn für u geringere Stetigkeitsvoraussetzungen gemacht werden. Wir betrachten daher die Integral-relation (96) als eine *Erweiterung der Differentialgleichung*. (Courant–Hilbert's italics.)[70]

Courant proved the following theorem about the generalized solutions:

Ist u eine zweimal stetige differenzierbare Lösung mit Ausnahme einer zweimal stetig differenzierbaren Fläche $\varphi(x_1, \ldots, x_n) = 0$ längs deren die ersten herausführenden Ableitungen von u sprunghafte Unstetigkeiten haben dürfen, so muss $\varphi = 0$ charakteristisch sein, d.h. die Relation $\sum_{i,k=1}^{n} a_{ik}\varphi_{x_i}\varphi_{x_k} = 0$ befriedigen.

This is all that Courant did concerning generalized solutions. In the rest of the work he treated only piecewise regular solutions, for which class the above theorem need not hold. Therefore this inclusion of a definition of generalized solutions in a famous textbook in no way signals an established theory. A paradigm in the theory of generalized solutions to partial differential equations was not established until Schwartz' book [1950/51].

50. The use of generalized solutions in the theory of hyperbolic equations appears to have been less coherent than in potential theory. Again, this may be explained by the greater diversity of the field and its weaker exploration (the same reasons given in §42 for the late use of generalized solutions to hyperbolic equations).

There is one very important person missing from this part, namely Sobolev. His work concentrated on the solution of the Cauchy problem, so that he would fit very nicely into this part. However, because of the outstanding nature of his work, I have chosen to discuss it in a separate part at the end of this chapter. This is to indicate that Sobolev was the one who carried furthest the development of the generalized solutions to partial differential equations before L. Schwartz. Thus Sobolev has unfortunately been left out of the chronology. His work belongs around 1935/36, before both Courant–Hilbert and Bochner and before much of the theory that shall be discussed in the next part.

Part 7. Differential Operators in Hilbert Spaces

51. The application of the abstract theory of operators to concrete differential operators gave rise to yet another generalization of differential operators.

In [1927] and [1930] von Neumann gave the axiomatic definition of a (separable) Hilbert space[70a] to be used in his new version of quantum mechanics (see Ch. 4, §30). In the same connection he developed the theory of (unbounded) operators in Hilbert spaces, primarily the spectral theory which is of importance in the determination of stationary states. However, in spite of its physical origin, the abstract operator theory was not directly applicable to those differential operators which were of interest in applications, quantum mechanics for one. For example, von Neumann had proved the spectral theorem for self-adjoint operators. However differential operators are not self-adjoint in their natural domain although they are equal to their formal adjoint; their domain is too small.[71]

In 1934 Kurt Friedrichs (born 1901) provided a method for extending a semibounded[72] symmetric operator to a self-adjoint operator.[72a] In this way he was able to use the abstract spectral theorem on the Schrödinger operator $-\Delta + v$ which is lower semibounded for suitable potentials v. The extension was made along the following lines [Friedrichs 1934, p. 480, note 14]. Friedrichs first considered the bilinear form (f, Ag) defined in the range $D(A)$ of A. He extended this [theorem 3] to a closed bilinear form by completion of $D(A)$ equipped with the scalar product

$$(f, Ag). \tag{97}$$

He found a maximal operator A' in terms of which this completed form was given by

$$(f, A'g). \tag{98}$$

This operator A' was a self-adjoint extension of A as Friedrichs showed [theorem 9].

52. Some mathematicians were not satisfied with this way of finding the self-adjoint extension. Even Friedrichs himself raised the question of an alternative method: "Offen bleibt aber noch, ob eine solche Fortsetzung auch auf andere Weise möglich ist." I. Halperin [1937] gave such an alternative construction which used only differentiation a.e. in the sense of Lebesgue in $\mathbb{R}^1$.

> Other writers have been able to avoid the real variable difficulties of determining the closure and adjoint by using a different inner product [i.e. 97], but it is desirable to consider the operator in classical Hilbert space.[73] [Halperin 1937.]

In [1939] Friedrichs himself gave another and (according to him [footnote 11]) better method for constructing the self-adjoint extension of the Schrödinger operator. It was based on the observation that an operator A^*A is self-adjoint when A^* is the adjoint of A:

> This suggests considering the above operator [the Laplacian] as the product D^*D of the gradient D . . . , and the negative divergence D^*.

where D and D^* are operators between two different Hilbert spaces. Thus Friedrichs' idea was to generalize the gradient and divergence operator so that they become the adjoints of each other.

> For the operator D this extension is possible by permitting differentiability in the sense of Lebesgue's theory. A correspondent definition of D^* does not seem to be

obvious. Therefore, it seems preferable to effect the extension of both operators in the following different way giving no preference to D or D^*.

First we consider the operator D applied on functions which have continuous derivatives and vanish in the neighbourhood of the boundary of S. We then define D in G_∞ as the closure and D^* in G' as the adjoint of this operator. We say: D in G_∞ is defined in the strong and D^* in G' in the weak sense.

Secondly we consider the operator D^*, applied on systems of functions with continuous derivatives and define D^* in G' as the closure (strong sense), and D in G_∞ as the adjoint (weak sense) of this operator.

Our main theorem is that the strong and the weak extensions coincide.

The main theorem allowed Friedrichs to apply the spectral theorem to the function DD^*.

Friedrichs' strong definition is a sequence definition: a function f is an element of the domain of the strong extension of $D(D^*)$ if there exists a sequence of functions f_n in the original domain and an L^2 function f_1 such that

$$\lim_{n\to\infty} \|f_n - f\|_2 + \|D^{(*)}f_n - f_1\|_2 = 0. \tag{99}$$

$D(f)(D^*(f))$ is then defined to be $f_1 = \lim D^{(*)}f_n$ in L^2.

It is clear from the definition of the adjoint operator (note 71) that the weak definition is a test function definition. In the process of defining the more elaborate extensions Friedrichs even gives another weak extension of D in $C_c^1(\Omega)$ which is closer to the one used in the theory of distributions: a function u is an element of the domain of the weak extension of D if there exists a system u' of $L^2_{\text{loc}}(\Omega)$ functions such that

$$\int_\Omega w'(x)u'(x)\,dx = \int_\Omega D^*w'(x)\cdot u(x)\,dx \tag{100}$$

for all systems $w'(x)$ in $C_c^1(\Omega)$. In this case $D(u)$ is defined to be equal to u'. The weak extension of D^* is defined in a similar way.

In a later paper [1944] Friedrichs proved the equivalence of the weak and strong extensions of any first-order differential operator with C^1 coefficients.

Thus his establishment of the equivalence of the weak and the strong extensions corresponds to the theorems proved earlier by Wiener (§43), Leray (§46) and Sobolev (§59) and proved later by Schwartz and Bochner (§48). His theorem differs from that of his predecessors and successors only in the different types of convergence used in the definition of the sequence extension and in the different test functions used in the test function definition.[74]

53. At the beginning of the previous quotation, Friedrichs mentioned the possibility of an extension of D using Lebesgue's theory. He probably meant an extension similar to Evans' generalization. Friedrichs rejected this procedure. However, I have already mentioned (§35, 36) another mathematician, J. W. Calkin, who developed Evans' theory with special reference to "the study of partial differential equations by means of the theory of transformations in Hilbert space". M. Stone had directed Calkin to this problem by suggesting as his doctoral thesis "Applications of the theory of Hilbert space to partial differential equations" (cf. [Calkin 1939]). A part of this thesis (1937) was published in a revised version in 1939 [Calkin 1939]. In this paper, however, Calkin only treated abstract operator theory: "A fifth chapter dealing with the applications of the theory to certain types of differential operators was originally planned, but is not included; applications will be considered in a subsequent paper."

The subsequent paper "Abstract definite boundary value problems" was published in [1940b]. Even though he and Morrey had developed their theory of generalized differential operators by then (§35, 36), Calkin did not build on these ideas. Instead of the extensions of the differential operators to the spaces $\mathfrak{P}$ and $\mathfrak{P}_\alpha$, he used a method similar to Friedrich's and Stone's, which he had treated in [1939] and [1940c]. Nowhere else in Calkin's limited production [Calkin 1940b, 1941] after 1940 did he use his and Morrey's method of extension. Apparently he had come to realize that Friedrich's and Stone's approach was superior to his and Morrey's.[75]

Thus Calkin and Morrey's method of generalizing differential operators never yielded new results in the Hilbert space theory though Calkin's contribution was motivated by the theory of differential operators in Hilbert spaces.

Part 8. Sobolev's Functionals

54. Sergej L'vovic Sobolev (born 1908), after being trained in the classical St. Petersburg school, devoted most of his life to a profound study of partial differential equations. His special interest in the wave equation and other hyperbolic equations reflects his employment at the Seismological Institute during the first three years (1929–1932) after he had done his diploma work. In 1933 Sobolev began a series of papers on the Cauchy problem. This led him two years later to that generalization of the function concept which makes his work so interesting from our point of view.[76] His treatment of the Cauchy problem of a normal hyperbolic differential equation continued the standard work of Hadamard [1908, 1932] (see Ch. 4, §8 and §11–13), but followed new and simpler lines, reducing the problem to an integral equation which could be solved by successive approximations. In a short note, "Le problème de Cauchy dans l'espace des fonctionelles" [1935b], Sobolev sketched how a very general existence theorem for the Cauchy problem could be obtained by extending the problem to a certain space of functionals. The note had no proofs and, according to the reviewer in the *Jahrbuch über die Fortschritte der Mathematik*, was "zu knapp", but Sobolev held out the prospect of a more detailed account of the matter. This appeared the following year in the article, "Méthode nouvelle à résoudre le problème de Cauchy pour les équations linéaires hyperboliques normales" [1936a], which I shall now examine.

55. The task Sobolev set himself in the paper was the solution of the equation

$$Lu \equiv \sum_{i=1}^{2k+1} \sum_{j=1}^{2k+1} A_{ij} \frac{\partial^2 u}{\partial x_i \, \partial x_j} + \sum_{i=1}^{2k+1} B_i \frac{\partial u}{\partial x_i} + Cu - \frac{\partial^2 u}{\partial t^2} = F, \tag{101}$$

with the initial conditions

$$\begin{aligned} u|_{t=0} &= u^{(0)}(x_1, x_2, \ldots, x_{2k+1}), \\ \left.\frac{\partial u}{\partial t}\right|_{t=0} &= u^{(1)}(x_1, x_2, \ldots, x_{2k+1}). \end{aligned} \tag{102}$$

Here A_{ij}, B_i and C are analytic functions of $x_1, \ldots, x_{2k+1}$ and t, and the quadratic form $\sum_{i=1}^{2k+1} \sum_{j=1}^{2k+1} A_{ij} p_i p_j$ is positive definite (which ensures that L is hyperbolic).

Under the explicit assumption that a solution u to (101) and (102) exists with bounded derivatives up to order $k + 1$ inclusive, Sobolev used a procedure, which he had developed in [1934] to find a formula similar to that of Kirchhoff (Ch. 4, §6–7), which in fact was an integral equation in u. Successive approximations gave him an explicit expression for u:

$$u(x_1, \ldots, x_n, t) = \sum_{n=0}^{\infty} J_n, \tag{103}$$

where the J_0 was uniquely determined from $u^{(0)}$, $u^{(1)}$ and F, and the J_ns were defined recursively from J_0. The sum (103) was shown to converge under the conditions set on u.

From a purely mathematical point of view this solution is not completely satisfactory because:

(1) the natural condition (classically) to impose on u would be twice (continuously or bounded) differentiability; and
(2) in order to obtain existence theorems, the assumptions should not be made about u but about the given quantities $u^{(0)}$, $u^{(1)}$ and F.

To obtain such an existence theorem Sobolev generalized the problem to a space of functionals.

56. The basic function spaces Φ_s were defined to consist of the s-times continuously differentiable functions of the $2k + 2$ variables $x_1, \ldots, x_{2k+1}, t$, with compact support[77] (i.e. $\Phi_s = C_c^s$) Sobolev defined convergence in Φ_s: φ_n is said to converge to φ in Φ_s if:

(1) the supports of all the φ_ns are contained in one bounded set;
(2) φ_n and its derivatives of order less than or equal to s converge uniformly to φ and its corresponding derivatives.

Sobolev wrote

$$\varphi_n \overset{s}{\Rightarrow} \varphi$$

(For $s = \infty$ this is precisely the convergence introduced by Schwartz in $\mathscr{D}\,(\sim C_c^\infty)$.)

The fundamental spaces of functionals Z_s were then defined to be spaces of the linear functionals on Φ_s which are sequentially continuous with respect to the above convergence. In Z_s Sobolev also introduced a notion of convergence $\rho_n \xrightarrow{s} \rho$, namely the weak convergence defined by

$$\rho_n \xrightarrow{s} \rho \Leftrightarrow (\rho_n, \varphi) \to (\rho, \varphi) \qquad \forall \varphi \in \Phi_s. \tag{104}$$

(This differs from Schwartz' notion in that Schwartz required the convergence on the right-hand side of (104) to be uniform on any bounded set of φs—the strong convergence.)

The functions ρ, integrable with their first l derivatives in every bounded subspace of $\mathbb{R}^{2k+2}$ defined special simple functionals in Z_s ($s = 0, 1, 2, \dots$), called the functionals of degree l, by the formula

$$(\rho \cdot \varphi) = \iiint \rho(M)\varphi(M)\, dR_{2k+2}, \qquad \forall \varphi \in \Phi_s. \tag{105}$$

(Schwartz used the same formula to imbed the space L^1_{loc}, which contains all Sobolev's functions ρ, in the space of distributions.)

Sobolev showed that the functionals of infinite degree were sequentially dense in Z_s. His proof, which has now become classic, consisted in showing that the functional (105) determined from the C^∞ function

$$\rho_\eta(M) = (\rho \cdot \omega_\eta^M), \tag{106}$$

where ω_η^M is a suitable C_c^∞ function of unit integral with support in an η ball around M, converges in Z_s to ρ for $\eta \to 0$. (ω_η^M is an approximation of $\delta(x - M)$.) (Schwartz proved the slightly stronger theorem that $\mathscr{D}$ is dense in $\mathscr{D}'$ [Schwartz 1950/51, p. 76].)

For a continuous linear operator $L\colon \Phi_{s_2} \rightharpoonup \Phi_{s_1}$ Sobolev showed that an adjoint operator

$$L^*\colon Z_{s_1} \rightharpoonup Z_{s_2} \tag{107}$$

was uniquely defined by the requirement that:

$$(L^*\rho, \varphi) = (\rho, L\varphi), \qquad \forall \rho \in Z_{s_1},\ \phi \in \Phi_{s_2}. \tag{108}$$

He used this procedure to define multiplication with a $C^{s'}$ function in $Z_{\min(s', s_2)}$ as the adjoint of multiplication in Φ_{s_2}, and differentiation $\partial/\partial x_i$ in $Z_{s_2 - 1}$ as the adjoint of $-\partial/\partial x_i$ in Φ_{s_2}. He proved (by partial integration) that the new operations were consistent with the old ones when applied to a functional of degree l (105). Thus the operator L (101) is generalized to apply in the spaces Z_s. (This generalization of differential operators is identical with the method later used by Schwartz, complicated only by the use of different values of s in Sobolev's theory.)

57. After these functional analytical definitions of the functional spaces, Sobolev returned to the original problem (101) and (102). He defined $\underline{u}$ to be the functional of degree 0 corresponding to the function which is equal to u for $t > 0$ and zero for $t < 0$, i.e.

$$(\underline{u} \cdot \varphi) = \iiint \underline{u}\varphi \, dR_{2k+2} = \iiint_{t>0} u\varphi \, dR_{2k+2} \tag{109}$$

and showed that if u satisfied (101) and (102) then $\underline{u}$ would satisfy

$$\left.\begin{aligned} L\underline{u} &= \rho, \\ \underline{u}|_{t<0} &= 0, \end{aligned}\right\} \text{in } Z_s, \tag{110, 111}$$

where ρ is the functional defined by

$$(\rho \cdot \varphi) = \iiint_{t>0} F\varphi \, dR_{2k+2} + \iint_{t=0} \left(u^{(0)} \frac{\partial \varphi}{\partial t} - u^{(1)}\varphi\right) dR_{2k+1} \tag{112}$$

and $\underline{u}|_{t<0}$ is said to be equal to zero when $(\underline{u}, \varphi)$ does not depend on the values of φ for $t < 0$, (in the language of distributions: $\text{supp}(u) \subset \{M | t \geq 0\}$). (Here F is the right-hand side of (101).)

Conversely if $\underline{u}$ is a solution to (110) and (111) which corresponds by formula (109) to a function u with integrable derivatives of first and second order then Sobolev easily showed that u is an ordinary solution to the original Cauchy problem (101) and (102).

Therefore it was natural for Sobolev to seek a solution to (110), i.e. to look for an inverse of the operator L. However, he did not construct the inverse on the whole space Z_s, but on restricted subspaces. These were defined as follows.

A set in $\mathbb{R}_{2k+2}$ is called a direct hyperbolic domain if its intersection with any direct characteristic cone of L (i.e. having t maximum at the vertex) is bounded. The direct hyperbolic space Ψ_s is defined to consist of functions $\psi \in C^s$ whose support is a direct hyperbolic domain. Introducing a convergence in Ψ_s similar to the one in Z_s, Sobolev considered the space W_s of continuous linear functionals on Ψ_s.

Similarly Ω_s is the space of C^s-functions with support in some inverse characteristic cone (for which t assumes its minimum value at the vertex), and the continuous (in an obvious sense) linear functionals on Ω_s constitute the space Y_s. Obviously

$$\Phi_s \subset \Omega_s \subset \Psi_s \tag{113}$$

and

$$Z_s \supset Y_s \supset W_s. \tag{114}$$

Sobolev remarked that Y_s consisted of all functionals from Z_s with support in a direct hyperbolic domain and W_s of all Z_s-functionals with support in a direct characteristic conoid.

Using the explicit expression for the solution (103) and functional analytical arguments, Sobolev showed that the operator L defined in the spaces $Y = \bigcup_{s=0}^{\infty} Y_s$ and $W = \bigcup_{s=0}^{\infty} W_s$ had a right-hand and left-hand inverse G.

The inversion theorem in Y can be used to prove the existence and uniqueness of a solution to (110) since both ρ and the desired $\mathbf{u}$ lie in Y. Since G is a right-hand inverse, $G\rho$ is seen to be a solution, and since G is a left-hand inverse, application of G to (110) shows that it is a unique solution.

This is Sobolev's beautiful existence and uniqueness theorem. From the remarks following (112) Sobolev concluded that if this functional solution was given by a C^2 function by (109), then this solves the classical Cauchy problem (110), (111) and if the solution is not of that form, the Cauchy problem has no classical solution. One instance in which Sobolev could prove the existence of a classical solution was when $F \in C^{k+1}$, $u^{(0)} \in C^{k+3}$ and $u^{(1)} \in C^{k+2}$.

58. Let us compare Sobolev's achievement with Schwartz' later theory of distributions. The points of resemblance are easily seen: Sobolev had the same aim as Schwartz, namely to generalize the function concept and some classical operations with functions to a larger field in which a certain problem, the Cauchy problem, could be solved more easily. The method of generalization of both the concept of function and of the operations was on the whole the same, using continuous linear functionals on certain function spaces and adjoint operators; even the notion of convergence in the function spaces coincided in the two expositions. Sobolev invented several spaces of functionals: the Z_s, Y_s and W_s for $s = 0, 1, \ldots$ and perceived the relationship between them both set-theoretically and topologically.

Thus Sobolev was the inventor of distributions, although the name "distributions" was only given later to the functionals by Schwartz.

However, there are obvious differences between Sobolev's and Schwartz' theories; I have already noted several of the discrepancies between the definitions and theorems found in Sobolev's paper and the corresponding statements in Schwartz' exposition. It should be added that the spaces Y_s and W_s have no parallel in Schwartz' theory. The most essential difference however is with the concepts, notions and theorems which are treated in Schwartz' book, but which cannot be found in Sobolev's article: for example, the spaces $\mathscr{E}'$ and $\mathscr{S}'$ (the distributions with compact support and the temperate distributions), the concepts of Fourier and Laplace transformation, tensor products and convolution, and the introduction of the δ-distribution and the partie finie. Sobolev defined only those spaces (Z_s, W_s, Y_s) and those concepts: differential operator, convergence and support (although not under that name), which were used in the treatment of the Cauchy problem. Here we find the kernel of the difference between Sobolev's and Schwartz' theories of distribution. Sobolev's was only a tool for the solution of one specific problem, whereas Schwartz developed the theory of distributions into a versatile theory, applicable and applied to the accurate statement and solution of many problems. With respect to the different lines in the prehistory of distributions, the difference can be stated thus: Sobolev fits into only one of these lines, namely that of generalized solutions to differential equations, whereas Schwartz tied all of the different trends together.

Thus Sobolev invented distributions, but it was Schwartz who created the theory of distributions.

59. Sobolev did not pursue the study of the spaces of functionals. In his continued study of differential equations, he kept to the traditional concept of a function (see, for example, his famous textbooks [1963] and [1964]), but used a generalized notion of differentiation. His ideas about generalized derivatives had actually preceded the theory of functionals by some months (cf. Ljusternik and Visik [1959, p. 206]). In an article on the diffraction of waves on Riemann surfaces [1935a] Sobolev defined a generalized (or limiting) solution in a domain Ω of the wave equation

$$\frac{\partial^2 u}{\partial x^2} + \frac{\partial^2 u}{\partial y^2} = \frac{1}{c^2}\frac{\partial^2 u}{\partial t^2} \tag{115}$$

as a function u for which a sequence $u_n \in C^2$ of ordinary solutions existed which converged in $L^1(x, y, t)$ to u. He showed in the same paper that a necessary and sufficient condition for u to be a generalized solution to (115) in this sense is that

$$\int u\left(\frac{\partial^2 v}{\partial x^2} + \frac{\partial^2 v}{\partial y^2} - \frac{1}{c}\frac{\partial^2 v}{\partial t^2}\right) dx\, dy\, dt = 0 \tag{116}$$

for all $v \in C^2(\Omega \cup \partial\Omega)$ with v and its first derivatives vanishing on the boundary of Ω.

Sobolev thus began with a sequence definition of a generalized solution and proved it was equivalent to a test function definition, a definition which he developed further during the same year into the definition of the generalized functions.[78]

In his later work he usually defined generalized solutions by means of test functions, but he occasionally used the sequence definition. Even in his very influential work on the spaces W_p^ν, which were later named after him, and which I will now investigate briefly, he did not use the theory of functionals.

60. Sobolev's work on the Sobolev spaces W_p^ν (or H_p^ν) was (according to Ljusternik and Visik [1959]) motivated by his variational treatment of the polyharmonic equation [Sobolev 1936c]. Two theorems, which anticipated the invention of the Sobolev spaces, were singled out in a separate article [1936b]. The theorems stated:

> Considerons l'ensemble de toutes les fonctions $\varphi(x_1, x_2, \ldots, x_n)$ continues dans un domaine fermé D de l'espace, ayant des dérivées continues jusqu'à l'ordre s à l'intérieur de ce domaine.
>
> Désignons par $L_s(A)$ la famille des fonctions de cet ensemble qui satisfont aux inégalités suivantes.
>
> $$\overbrace{\iint\cdots\int}^{n}\left(\frac{\partial^j \varphi}{\partial x_1^{a_1}\,\partial x_2^{a_2}\cdots\partial x_n^{a_n}}\right)^2 dx^1\, dx^2 \cdots dx^n \le A;$$
>
> $$a_1 + a_2 + \cdots + a_n = j, j = 1, 2, \ldots, s. \tag{117}$$
>
> En faisant quelques hypothèses restrictives sur le domaine D, nous pouvons établir les deux theorèmes suivants:

Théorème I. *Les fonctions de la famille $L_k(A)$, où k est le plus petit nombre entier qui surpasse $n/2$, c.-à.-d. $k = [n/2] + 1$, sont bornées dans leur ensemble.*

Théorème II. *La famille $L_k(A)$ satisfait également à la condition de Hölder d'ordre $\alpha < 1$ pour n paire et d'ordre $\alpha = 1/2$ pour n impaire.*

In Sobolev [1936c] these two theorems were generalized to describe the behaviour of the mean value of the square of difference of the function on an $(n - s)$-dimensional submanifold and a neighbouring manifold. Kondrachow [1938] generalized this further from an L^2 statement to an L^p statement. But in all these works there were no generalized derivatives and no explicit mention of the Sobolev spaces.

In an article two years later [Sobolev 1938b] (brief exposition in Sobolev [1938a]) Sobolev spaces were defined explicitly with the aid of generalized derivatives introduced in the following "test function" way: if in a domain $D \subset \mathbb{R}^n$

$$\iint \cdots \int_D \left(\varphi \frac{\partial^\alpha \psi}{\partial x_1^{\alpha_1} \cdots x_n^{\alpha_n}} + (-1)^{\alpha - 1} \psi \omega \right) dx_1 \cdots dx_n = 0 \tag{118}$$

for all $\psi \in C_c^\alpha(D)$, then ω is the generalized derivative

$$\omega = \frac{\partial^\alpha \varphi}{\partial x_1^{\alpha_1} \cdots \partial x_n^{\alpha_n}}. \tag{119}$$

The space $L_p^{(\nu)}$ which is now usually called W_p^ν or H_p^ν was then defined as the space of functions whose generalized derivatives up to order ν are in L^p. Sobolev announced two imbedding theorems in [1938b].

(A) For $\nu > (n/p)$: $L_p^{(\nu)} \subset C^{\nu - [n/p] - 1}$ and the imbedding is sequentially continuous for a suitable convergence in $L_p^{(\nu)}$.

(B) For $\nu < (n/p)$: $L_p^{(\nu)} \subset L^{(\nu - l)}_{1/[(1/p) - (l/n)]}$.

He asserted that theorem A had been proved in his [1936b] article for the case when $p = 2$ and by Kondrachow [1938] for general p. Since the theorems in these papers are not formulated in terms of generalized derivatives, this statement is obviously not precise. However, a density argument easily gives A as a corollary to theorem I above and the corresponding L^p theorems.[79] Theorem B was proved in detail by Sobolev [1938b] by first proving its validity for the case when all derivatives were continuous and then passing to the limit, exactly as was done in (note 79) in the proof of A.

61. As has been mentioned several times in this chapter, Sobolev was not the first to define or use the Sobolev spaces. They had already appeared in many problems on partial differential equations, especially in the variational treatment of the Dirichlet problem, the problem which had motivated Sobolev himself. With different definitions of the generalized derivatives and different values of ν and p Sobolev spaces had been used by Beppo Levi [1906] (§24), Tonelli [1926a, b] (§20), Nikodym [1933a] (§25) and Evans [1920] (§33, 34).

Calkin [1940] and Morrey [1940] (§35) used them later independently of Sobolev.

Sobolev, who knew very well about some of his predecessors, wrote about Theorems I and II [1936b]:

> Nos théorèmes donnent une précision des évaluations connues, due à l'école de Göttingen, qu'on rencontre souvent dans differents problèmes de la théorie des équations aux dérivées partielles.[80]

After Sobolev's proof of the imbedding theorems Sobolev spaces became widely applied both by Sobolev himself and by other mathematicians (references can be found in Morrey [1964]). Their importance lies in the fact that they allow one to conclude, in many instances, that a generalized solution to a partial differential equation belonging to a Sobolev space is in fact an ordinary solution. The imbedding theorems were also strengthened by Friedrichs [1944] (see note 74) and Kryllov [1947].

Thus generalized differentiation came to play a large role in the theory of Sobolev spaces, but Sobolev's functionals (distributions) were not used in this connection until Schwartz had rediscovered them. Using distributions it has been possible to define the Sobolev spaces H_p^ν for negative ν; in modern texts the spaces H_2^ν are defined with the aid of the Fourier transformation for all real values of ν.[81]

62. In a note written on the occasion of Sobolev's fiftieth birthday, Ljusternik and Visik [1959] wrote:

> Here [in 1935b and 1936a] Sobolev has laid the foundation of the generalized functions, which has got such vast applications in modern analysis. Thus the further development of this theory took place within the frames of the concept indicated by Sobolev.
>
> In his works from 1935 Sobolev not only laid the foundation of the generalized functions, but also proved in an important example—namely the Cauchy problem for hyperbolic equations—the applicability of these concepts.

This aptly describes Sobolev's great achievements in the prehistory of the theory of distributions. It should be kept in mind, however, that although the further development of the theory took place within the framework indicated by Sobolev, it was not the work of Sobolev, but primarily of L. Schwartz.

According to Schwartz [1974, 1950/51], he did not know of Sobolev's work before 1945.

Part 9. Methods. A Survey

63. In this section I shall summarize briefly the main ideas in this chapter on generalized differentiation and generalized solutions to partial differential equations. Instead of ordering the material according to the fields of applications, as has been done previously, I will focus here on the methods used in the generalizations. Hence this section will not include new material. The survey will consist of a list of the different generalization methods we have seen in the previous sections together with mention of their characteristics, their applications, and the persons who applied them, together with references to the relevant pages in the previous sections.

A. Substitution of the Differential Equation with Another Model of the Physical System, Based on Physical Arguments

The alternative model is aimed at treating singularities with which the original model cannot deal. From a mathematical point of view such a generalization is not very interesting whereas from a physical point of view it is more satisfactory than the mathematical methods (Ch. 2, note 8).

The method was applied to equations of motions—hyperbolic partial differential equations—by Lagrange 1759 (§11), Riemann 1859 (§14), Christoffel 1876 (§14), Harnack 1887 (note 14), Oseen 1911 (§45) and many others.

B. Space Geometric Method

Partial derivatives (e.g. $\partial f(x, y)/\partial x$ and $\partial f(x, y)/\partial y$ are conceived of as geometrical objects which can be found by intersecting the surface $z = f(x, y)$ with given planes parallel to the z axis (in the example, parallel with the x and y axis, respectively). If the intersections are equal, the partial derivatives are considered to be equal (in the example $\partial f/\partial x = \partial f/\partial y$) (cf. the figure in note 12). The method conceals the problem: Which values do the partial derivatives have at the irregular points?

This method was used for all types of partial differential equations in two independent variables by Monge in the 1770s (§12) and Arbogast 1787 (§13).

C. Substitution of Many Limits with One

Higher-order derivatives can be defined by taking one limit instead of taking the successive limits in all the lower-order derivatives. This was done by Lagrange 1760/61 (§11) and Riemann 1867 (§16) in the theory of the vibrating string and the theory of Fourier series, respectively. Compound differential expressions can be considered as a whole instead of as a sum of separate operators each of which involve a limiting procedure. This variation of the method was used in potential theory by Petrini 1908 (§28) and in a certain sense by Wiener 1927 (§41). Evans suggested 1920 (§31) combining the limits in a sum of second-order derivatives into one and thus used both versions of this method at the same time.

Substitution of an ordinary limit with a lim sup(inf.) was used by Dini in 1878 (§19).

D. Differentiation a.e. of Absolutely Continuous Functions

In one dimension the generalization was made in connection with measure and integral theory (the fundamental theorem of the calculus) by Lebesgue 1902 and Vitali 1905 (§19) and had been suggested earlier in a different form by Harnack 1882 (Riemann measure) (note 24).

In two dimensions it was applied in measure-and-integral-theory by Tonelli 1926 (§20) and Morrey (§20) and in the calculus of variation by Beppo Levi 1906 (§24), Fubini 1907 (§24) Tonelli 1929 (§24, note 37) and Nikodym 1933 (§25). In the variational calculus the absolute continuity was always combined with the assumption that the generalized derivatives must be in a certain L^p space (Sobolev spaces).

In Hilbert space theory the method was applied by Murray 1935 and Halperin 1937 (§52, note 73).

E. Test Curves and Test Surfaces

Assuming one of the functions in Green's theorem (or another of the main theorems of vector analysis) equal to 1, the given differential equation can be transformed into an integro-differential equation, which must hold for all sufficiently regular domains of integration, which may be either curves or surfaces (test curves, test surfaces).

The new equation has meaning for a larger group of functions than the original equation and in this way generalizes it (see §30).

The method was mainly applied in potential theory by Bôcher 1905/06 (§29), Weyl 1913 and 1940 (§37, 39), Evans 1914–1933 (§31–34), Calkin 1940 (§35–36) and in a certain sense Wiener 1927 (§41). We have encountered one application to parabolic equations: Evans 1914 (§32) and one to the calculus of variations: Morrey 1940 (§35–36).

Calkin's and Morrey's work was closely linked with method *D*.

F. Test Functions

The differential equation is multiplied with a sufficiently regular test function which has compact support in a given domain and the product is integrated partially over this domain. In this way the differential operator is transferred to the test function as the adjoint operator so that the resulting integro-differential equation, which is supposed to hold for all test functions, does not presuppose differentiability of the solution. The method is closely connected to method *E*. Instead of choosing one of the functions in Green's theorem (which gives a type of partial integration) equal to 1, one considers it to be an arbitrary test function with compact support in a fixed set, which does not vary as in method E.

In Hilbert space terminology this generalization is called the weak extension.

The test function method arose in the theory of hyperbolic (and parabolic) partial differential equations (the Cauchy problem) where it was anticipated by Lagrange 1761 (§11), used in an incomplete way by Harnack 1887 (note 14) and explicitly formulated and applied by Wiener 1926 (§43) (operational calculus), Leray 1934 (§46), Sobolev (§54–62) and Courant and Hilbert 1937 (§49). It was soon applied to other partial differential equations by Friedrichs 1939 (§52), Weyl 1940 (§38), Schwartz 1945 (Ch. 6) and Bochner and Martin 1948 (§48).

This method gave the basis for the theory of distributions.

G. The Sequence Method

According to this method, a function is said to be a generalized solution to a given differential equation if there exists a sequence of classical solutions which converges to the function. The type of convergence varies from author to author. In Hilbert space terminology (L^2 convergence) this generalization is called the strong extension.

The sequence definition was anticipated by Euler 1765 (§10*c*) and Laplace 1772 (§12, note 10). In connection with method *D* it was used as a definition by Lewis 1933 (§47). In its pure form it was used for the first time as a definition by Sobolev (L^1 convergence) 1935 (§59) in his treatment of the Cauchy problem, and later used by Friedrichs (convergence in the norm $\|f\| + \|Df\|$) 1939 (§52), L. Schwartz (uniform convergence on compact subsets) 1944 (Ch. 6, §4), and Bochner (weak convergence in a neighbourhood of each point) 1945 (§48). In addition, the sequence property was used, but not as a definition, by Wiener (L^2 convergence) 1926 (§43), and Leray (weak or strong L^2 convergence) 1934 (§46).[82]

64. The preceding seven methods are obviously not independent. In the previous sections we have seen how similarities or differences were pointed out by some of the mathematicians involved. I will shortly summarize these connections.

$A \leftrightarrow B \leftrightarrow C$

The connection between the first three methods, and to some degree *F* and *G* also, was debated in the early discussion of the vibrating string (§5–13). The discussion was based on concepts which have now been abandoned, but portions of the argument can be translated into modern terms. Nevertheless, we cannot derive information from it which is sufficiently precise to be of any mathematical value today.

$D \leftrightarrow E$

Evans [1933] proved that Tonelli's version of *D* was a special case of his own definition *E*. In Calkin and Morrey [1940] the two methods were merged.

They used the term absolute continuity as in D, but the generality of their theory was the same as Evans'.

$D, E \leftrightarrow G$

Weyl showed in 1940 (§39) that test surface generalizations of the equations div $f = 0$, rot $f = 0$ implied test function generalizations of the same equations; in 1944 Friedrichs remarked that Bôcher's and Evans' generalizations "and our weak extension are related", but he did not spell out the relationship in detail.

$F \leftrightarrow G$

These two methods were developed in close connection with each other and can often be found in the same papers. The equivalence of the methods was proved by Wiener, Leray, Friedrichs, Sobolev, Bochner and Schwartz (see references under F and G). Wiener and Leray needed the sequence property but nevertheless chose to use the test function definition. On the other hand, Sobolev and Schwartz used sequence definitions before they made the test function method the basis of the theory of generalized functions.

$D \leftrightarrow F$

Sobolev pointed out in [1938a] that differentiability a.e. and differentiability defined with the aid of test functions are independent notions in the sense that: (1) if $\varphi = f_1(x_1) + f_2(x_2)$ then $\partial^2\varphi/\partial x_1\,\partial x_2 = 0$ in the test function sense while φ might not be differentiable a.e.; and (2) if $\varphi = F(x_1)$ is of bounded variation, but not absolutely continuous, then $(\partial/\partial x_1)\varphi$ exists a.e. but it does not exist in the test function sense.

On the other hand, Schwartz [1950, p. 58, Theorem V] showed that the existence of the distribution derivative $(\partial/\partial x_i)f$ as an L^1_{loc} function is equivalent to the absolute continuity of f on a.a. lines parallel with the x_i axis.

65. However interesting it may be from a mathematical point of view to organize the history according to methods, historically it is more satisfactory to arrange the material in this chapter according to problems.

It was the problems, and not a desire to investigate methods, that stimulated research in the theory of generalized differentiation and generalized solutions to partial differential equations. Mathematicians usually built upon predecessors who had worked on the same problem, and rarely upon those who had applied the same method to different problems.

This claim seems to be partially unjustified when the two charts at the end of the book are compared. In the period before 1930 the reference arrows seem to remain more within the same column in the problem chart than in the method chart. Still, there are many more arrows between works using the same method than between works using different methods. This phenomenon is easily explained as a side effect from the vertical arrows in the problem chart. For, as we have seen there was a strong connection between methods and the problems in which they were used. This connection was probably due both to tradition and to the fact that certain problems naturally suggested certain methods. For example, in problems in which Lebesgue integrals were involved (variational calculus, areas of surfaces, etc.) it was natural to use differentiation a.e. whereas in the solution of boundary-value problems in potential theory, test curves and test areas were natural tools.

After 1930 the charts indicate several references from one type of problem to another. This clearly shows that mathematicians began to view their methods as part of a larger whole: "generalized derivatives and generalized solutions to partial differential equations". However, the methods largely continued to be *ad hoc* in a development which was prompted by specific problems. A unified *theory* was not created until L. Schwartz' theory of distributions.

Chapter 3

Generalized Fourier Transforms

1. La théorie de la série et de l'intégrale de Fourier a toujours introduit de grandes difficultés et nécessité un appareil mathématique important pour mettre au point les questions de convergence.... Pour l'intégrale de Fourier, l'introduction des distributions est inévitable, sous une forme directe ou camouflée,

wrote Schwartz in his historical introduction [Schwartz 1950/51].[1]

One of the great merits of the theory of distributions is the simple generalization it gives of the Fourier transformations. The need for such a generalization became urgent at the beginning of this century. However it was not only the realization of the *problem* of generalized Fourier transforms, but also the *methods* used in its solution which anticipated the theory of distributions:

> Les méthodes de Bochner, de Carleman (transformée de Fourier analytique), de Beurling (transformée de Fourier harmonique) sont très proches de nôtres. [Schwartz 1950/51, Vol. I, p. 7.]

Of these three, Bochner came the closest to the theory of distributions. He came so close that he thought that Schwartz' theory was only a slightly improved version of his own methods. Schwartz acknowledged the similarity of the two theories, calling Bochner's objects "distributions":

> Les "distributions" de Bochner sont, au fond, définies comme dérivées de fonctions continues n'ayant pas nécessairement de dérivée usuelle; notre théorème XXI du chapitre III exprime justement qu'une distribution est, localement, une dérivée d'une fonction continue. Il nous parait bien préférable d'avoir cette propriété plutôt comme théorème que comme définition. [Schwartz, 1950/51, Vol. I, pp. 7–8.]

I shall discuss Bochner's generalization and its precursers in the first part of this chapter (§2–18).

Carleman's and Beurling's generalizations of the Fourier transformation anticipated another theory of generalized functions, the theory of hyperfunctions. These methods are discussed in the latter part of this chapter (§20–23).

2. The problem which "inevitably" introduced distributions into the theory of the Fourier integral was the following: In order that the Fourier integrals

$$\int_0^\infty f(x) \cos \mu x \, dx, \qquad \int_0^\infty f(x) \sin \mu x \, dx, \quad \text{and} \quad \int_{-\infty}^\infty f(x) e^{ix\mu} \, dx$$

exist, the function f must decrease quickly enough at infinity. Such a restriction on f, however, is very inconvenient for the application to differential equations or in harmonic analysis. Therefore it is desirable to find generalizations of the Fourier transformation by which functions which are only bounded at infinity or even increasing can be transformed. It was these generalizations which introduced distributions in a camouflaged version.

Of course a function also has to satisfy local regularity conditions if its Fourier integral is to exist. But these conditions are not very restrictive. It is sufficient that the function can be expanded in a Fourier series [Hahn 1926] and as Dirichlet [1829, 1837] already showed, a function can be very irregular and still admit of a Fourier series expansion. Therefore the local conditions were not a subject of special concern in the theory of Fourier integrals, and they will be omitted in the subsequent account.

3. Fourier introduced the Fourier integrals

$$\varphi(\mu) = \int_0^\infty f(x) \cos \mu x \, dx \quad \text{and} \quad \psi(\mu) = \int_0^\infty f(x) \sin \mu x \, dx \tag{1}$$

and their inverses

$$f(x) = \frac{2}{\pi} \int_0^\infty \varphi(\mu) \cos \mu x \, d\mu \quad \text{and} \quad f(x) = \frac{2}{\pi} \int_0^\infty \psi(\mu) \sin \mu x \, d\mu; \quad \text{for } x > 0 \tag{2}$$

in his prize paper [Fourier 1811, publ. 1824/26] dealing with heat diffusion. He elaborated the theory of Fourier integrals in his main work *Théorie Analytique de la Chaleur* [1822, §342–362]. Formulas (1) and (2) he obtained from the Fourier series expansion of a periodic function by letting the interval of periodicity tend to infinity. A combination of (1) and (2) gives the Fourier integral theorem

$$f(x) = \frac{1}{\pi} \int_0^\infty \int_{-\infty}^\infty f(y) \cos \mu(y - x) \, dy \, d\mu. \tag{3}$$

for which Fourier gave an alternative proof involving the δ-function (Ch. 4, §18).

During the nineteenth century the theory of Fourier integrals only received minor attention compared to its sister discipline, the theory of Fourier series, which occupied a central position in analysis. Several "proofs" of the integral theorem (3) were given, but they were mostly byproducts of similar theorems for Fourier series [Burkhard 1914, Ch. 5]. At the beginning of the nineteenth century the proofs of (3) did not spell out explicitly what conditions f had to fulfil at infinity. Later in the century, when the conditions on f were stated, (3) was only proved for a very restricted class of fs. In [1910] Pringsheim gave an elaborate description of the work which had previously been done in the field. He found only one condition which was general enough, namely the following due to Harnack:

> $f(x)$ tends to zero at infinity and has in a neighbourhood of infinity an absolutely integrable derivative.

However Pringsheim remarked that this condition was unsatisfactory from a theoretical point of view:

> als es die Güldigkeit einer *Integral*-Formel von der *Stetigkeit*, ja sogar von der *Differenzierbarkeit* der zu integrirenden Funktion abhängig macht, mithin Beschränkungen einführt, die dem Wesen der Sache fremd sind. [Pringsheim 1910, p. 368.]

Therefore in 1910 he gave several other conditions under which the Fourier integral theorem was valid. One of these stated that $f(x)$ must converge monotonically to zero; the other conditions are too technical to discuss here [Pringsheim 1910, pp. 405–406].

4. In the same year Plancherel published his Habilitations-schrift in which he gave his famous extension of the Fourier integral to the space $L^2\ (0, \infty)$. In this space he could not use the ordinary cosine-transform (1) since the integral might not exist. But he could use the expression

$$\varphi(\mu) = \sqrt{\frac{2}{\pi}}\frac{d}{d\mu}\int_0^\infty \left(\int_0^\mu f(x)\cos xv\,dv\right)dx = \sqrt{\frac{2}{\pi}}\frac{d}{d\mu}\int_0^\infty f(x)\frac{\sin\mu x}{x}\,dx, \tag{4}$$

which reduces to (1) if differentiation under the integral sign is allowed (apart from the difference in the constant).

Plancherel showed that the function φ defined by (4) was in $L^2\ (0, \infty)$ and that the transformation (4) applied to φ would give f again, i.e.

$$f(x) = \sqrt{\frac{2}{\pi}}\frac{d}{dx}\int_0^\infty \varphi(\mu)\frac{\sin x\mu}{\mu}\,d\mu \quad \text{a.e.} \tag{4a}$$

[Plancherel 1910, 1913, 1915].

5. Plancherel's theorem is of the utmost importance in functional analysis. From our point of view however, the first really significant generalization of the Fourier transformation was made by Hans Hahn (1879–1934) in 1924 in a talk presented at the *Jahresversammlung der Deutschen Mathematiker Vereinigung in Innsbruck* [Hahn 1924] and repeated in *Acta Mathematica* two years later [Hahn 1926].

Hahn's idea was to use Plancherel's formula (4) without the $d/d\mu$ in front. He did not explicitly state that he got the idea from Plancherel, but since he referred to the above-mentioned articles of Plancherel, it seems most probable that he did. Thus Hahn considered the transforms:

$$\Phi(\mu) = \int_{-\infty}^{\infty} f(x) \frac{\sin \mu x}{x}\, dx, \tag{5}$$

$$\Psi(\mu) = \int_{-\infty}^{\infty} f(x) \frac{1 - \cos \mu x}{x}\, dx.^2 \tag{6}$$

If the functions φ and ψ in (1) exist, they have the integrals Φ and Ψ respectively. Therefore it is not surprising that Hahn found the inversion formula:

$$f(x) = \frac{1}{\pi}\left(\int_0^{\infty} \cos \mu x \, d\Phi(\mu) + \int_0^{\infty} \sin \mu x \, d\Psi(\mu)\right), \tag{7}$$

where all the integrals are improper Lebesgue–Stieltjes integrals (i.e. $\int_0^{\infty} \cdot = \lim_{\lambda\to\infty} \int_0^{\lambda} \cdot$). Hahn called (7) the Fourier–Stieltjes integral.

Because of the xs in the denominators of (5) and (6) these integrals exist for a much larger class of functions than the ordinary Fourier integrals (1). Hahn proved the existence of (5) and (6) and the correctness of the inversion formula (7) under the following conditions at infinity, either

(1) $|f(x)/x|$ is integrable at infinity; or
(2) $f(x)$ is a product of a periodic function and a function which at infinity is bounded and monotone; or
(3) $f^2(x)$ is integrable at infinity.

The second condition allowed Hahn to "Fourier transform" constant functions. If $f(x) = 1$, Hahn found [1926, §8] that

$$\Phi(\mu) = \begin{cases} 0 & \text{for } \mu = 0, \\ \pi & \text{for } \mu > 0, \end{cases} \qquad \Psi(\mu) = 0.$$

Hahn's generalized Fourier transformation is easy to compare with the Fourier transformation in the distribution sense. Take, for example, an even function f. In this case $\Psi \equiv 0$, and the distribution derivative of Ω

$$\frac{d}{d\mu}\Phi(\mu) \tag{8}$$

is precisely the Fourier transform of f in the distribution sense. If, for example, $f(x) \equiv 1$, as above, the Fourier transform in the distribution sense is $\pi\delta$, corresponding to the fact that $(d/d\mu)\Phi(\mu) = \pi\delta$.

Hahn was aware of this connection between Φ and φ when φ exists. We therefore see that he deliberately avoided generalized functions by altering the transformations to (5) and (6) and using the Stieltjes integral. In modern terms one can say that the Stieltjes integral (7) is given by the Radon measure

$d\Phi(\mu)$, so that Hahn's generalized Fourier transform is in essence a functional. It would be an overinterpretation, however, to attribute this way of thinking to Hahn who does not speak of measures, and still less of functionals.[3]

6. Independently[4] of Hahn, Norbert Wiener (1894–1964) found a similar generalization of the Fourier integral [Wiener 1925]. Contrary to Hahn, however, he was not interested in pointwise convergence but in what he called "limit almost in the mean":

$\varphi(x, y)$ is said to converge almost in the mean to f

(abbreviated l.a.m. $\varphi(x, y) = f(x)$) in $[a, b]$ if

$$\lim_{\xi \to \infty} \int_{\xi}^{\xi+h} \int_a^b [\varphi(x, y) - f(x)]^2 \, dx \, dy = 0, \tag{9}$$

i.e.

$$\lim_{\xi \to \infty} \int_{\xi}^{\xi+h} \|\varphi(x, y) - f(x)\|_2^2 \, dy = 0. \tag{10}$$

Instead of the classical Stieltjes integral, which he considered inadequate, he used the integral

$$\mathop{\mathrm{I}}_{0}^{A} \varphi(x) \, d\psi(x) \equiv \varphi(A)\psi(A) - \int_0^A \psi(x)\varphi'(x) \, dx, \tag{11}$$

which can formally be obtained from the ordinary Stieltjes integral by integration by parts. (Apart from a constant.)

The functions for which Wiener proved the Fourier inversion formula were the "nearly bounded" functions, that is the functions for which the integral

$$\int_{\xi}^{\xi+h} [f(x)]^2 \, dx \tag{12}$$

is bounded for all ξ, where h is a positive constant.

Wiener showed that for f nearly bounded the functions

$$\gamma(\alpha) = \lim_{T \to \infty} \frac{1}{\pi} \int_{-T}^{T} f(\lambda) \cdot \frac{\sin \alpha\lambda}{\lambda} \, d\lambda, \tag{13}$$

$$\delta(\alpha) = \lim_{T \to \infty} \frac{1}{\pi} \int_{-T}^{T} f(\lambda) \frac{e^{-\lambda} - \cos \alpha\lambda}{\lambda} \, d\lambda^{5} \tag{14}$$

exist, where lim means L^2 limit over every bounded interval. From these two generalized Fourier transforms Wiener obtained the inversion formula:

$$f(x) = \mathop{\text{l.a.m.}}_{T \to \infty} \left[\mathop{\mathrm{I}}_{0}^{T} \cos \alpha x \, d\gamma(\alpha) + \mathop{\mathrm{I}}_{0}^{T} \sin \alpha x \, d\delta(\alpha) \right] \tag{15}$$

over any bounded interval. His proof of (15) was very complicated, covering five full pages of long integrals.

7. Hahn's and Wiener's work with generalized Fourier transforms differed in motivations and scope. Hahn seems to have been motivated by the purely mathematical desire to combine Fourier integrals and Fourier series. For Wiener the generalized Fourier transform was only a part of his revolutionary work on generalized harmonic analysis. This theory, which not only included the two theories unified by Hahn but also H. Bohr's theory of almost periodic functions, was created in order to study the harmonic analysis of noise and of white light.[6] In his classical article, "Generalized harmonic analysis" [1930], he explained this physical motivation as follows:

> The two theories of harmonic analysis embodied in the classical Fourier series development and the theory of Plancherel do not exhaust the possibilities of harmonic analysis.... Neither is adequate for the treatment of a ray of white light which is supposed to endure for an infinite time. Nevertheless the physicists who first were faced with the problem of analysing white light into its components had to employ one or the other of these tools. Gouy accordingly represented white light by a Fourier series, the period of which he allowed to grow without limit, and by focussing his attention on the average values of the energies concerned he was able to arrive at results in agreement with experiments. Lord Rayleigh, on the other hand, achieved much the same purpose by using the Fourier integral, and what we now should call Plancheral's theorem. In both cases one is astonished by the skill with which the authors use clumsy and unsuitable tools to obtain the right results, and one is led to admire the unfailing heuristic insight of the true physicist.

Wiener replaced the clumsy methods of the physicists with the more elegant method of generalized harmonic analysis (see Wiener's own account of the history of harmonic analysis in Wiener [1938]).

8. The importance of the Stieltjes integral was realized by Wiener. He outlined the currently prevailing theory of white light showing that only the discontinuous spectrum was taken into account and remarked further:

> The chief reason for this is that any measure for a continuous spectral density becomes infinite at a spectral line, while any measure for the intensity of a spectral line becomes zero over the continuous spectrum.
>
> This is a difficulty, however, which has had to be faced in many other branches of mathematics and physics. Impulses and forces are treated side by side in mechanics although they have no common unit. We are familiar in potential theory with distributions of charge containing point, line, and surface distributions, as well as continuous volume distributions. The basic theory of all these problems is that of the Stieltjes integral.

Thus Wiener was able to see the connection between such seemingly diverse topics as generalized Fourier integrals and potential theory and found the link in their common use of the Stieltjes integral. Yet he was still far from anything resembling a theory of generalized functions. First of all, as long as the "Stieltjes measures" $d\varphi$ were inseparable from integrals, a calculus of these "generalized functions" was excluded. Secondly, Wiener did not see the connection between the generalized Fourier transformations and the generalized solutions of differential equations for which he possessed two other methods (Ch. 2, §41 and §43). He did not pursue his idea on the Stieltjes integral any further.

9. Until now we have seen how the domain of the Fourier transformation could be extended by formally integrating once under the integral sign. I shall call this the 1-transformation, using a notation introduced by Bochner [1932]. This trick can naturally be repeated, giving rise to even more generalized transformations. This was actually done by both Hahn and Wiener immediately after they had taken the first step described above.

Hahn presented his theory to the "Akademie der Wissenschaften" in Wien at the end of 1925. He gave the integral expression [1925]:

$$f(x) = \lim_{\mu\to\infty} \frac{1}{\pi} \int_0^\mu \left(\int_0^\lambda \left(\cos \tau x \frac{d^2\Phi_2(\tau)}{d\tau} + \sin \tau x \frac{d^2\Psi_2(\tau)}{d\tau} \right) \right) d\lambda, \tag{16}$$

Φ_2 and Ψ_2 being defined by

$$\Phi_2(\mu) = 2 \int_{-\infty}^{\infty} f(x) \frac{\sin^2(\mu/2)x}{x^2} dx = \int_{-\infty}^{\infty} f(x) \frac{1 - \cos \mu x}{x^2} dx \tag{17}$$

and

$$\Psi_2(\mu) = \int_{-\infty}^{\infty} f(x) \frac{\chi[-1, 1]\mu x - \sin \mu x}{x^2} dx, \tag{18}$$

where $\chi[-1, 1]$ is the characteristic function on $[-1, 1]$.[7] I shall call (17) and (18) the 2-transforms of f.

Here the "Stieltjes integral" (16) is to be understood in the following way:

$$\int_a^b f(x) \frac{d^2 g}{dx} = \lim_{n\to\infty} \sum_{i=1}^{n-1} f(x_i) \left(\frac{g(x_{i+1}) - g(x_i)}{x_{i+1} - x_i} - \frac{g(x_i) - g(x_{i-1})}{x_i - x_{i-1}} \right), \tag{19}$$

where $a = x_0 < x_1 < x_2 < \cdots < x_n = b$ is a partition of the interval $[a, b]$.

Hahn was able to prove the existence of the integrals (17) and (18) and the identity (16) under the assumptions that f be continuous in x and bounded at infinity.[8] This is a less restrictive condition than the one given by Hahn the preceding year for the 1-transformation, but more restrictive than Wiener's condition (§6).

10. Hahn sent a draft of the talk to Wiener who immediately published his own research on similar integrals [1926a]. Here again Wiener differed from Hahn in his interest in "limit almost in the mean" (9) instead of Hahn's pointwise "Fejer" summation (16). In addition, instead of Hahn's definition of the "Stieltjes integral", he defined it with the aid of partial integration as in [1925].

More precisely Wiener showed that if f is a function for which

$$\lim_{A\to\infty} \frac{1}{2A} \int_{-A}^{A} [f(t)]^2 \, dt \tag{20}$$

exists then the following integrals exist:

$$\Gamma(\mu) = \frac{1}{\pi} \int_{-\infty}^{\infty} f(t) \frac{1 - \cos \mu t}{t^2} \, dt, \tag{21}$$

$$\Delta(\mu) = \frac{1}{\pi} \int_{-\infty}^{\infty} f(t) \frac{\mu t e^{-t^2} - \sin \mu t}{t^2} \, dt. \tag{22}$$

Moreover, since Γ and Δ are absolutely continuous, Wiener could define

$$\begin{aligned} \int_0^A \cos \mu t \, d\Gamma'(\mu) + \int_0^A \sin \mu t \, d\Delta'(\mu) \equiv {} & \cos(At)\Gamma'(A) \\ & + \sin(At)\Delta'(A) \\ & + t[\sin(At)\Gamma(A) - \cos(At)\Delta(A)] \\ & - t^2 \int_0^A \cos \mu t \Gamma(\mu) \, d\mu \\ & + \int_0^A \sin(\mu t)\Delta(\mu) \, d\mu \end{aligned}$$

(formally: partially integrating twice) and show that

$$\operatorname*{l.a.m.}_{A \to \infty} \left(\int_0^A \cos \mu t \, d\Gamma'(\mu) + \int_0^A \sin \mu t \, d\Delta'(\mu) \right) = f(t). \tag{23}$$

11. Wiener's condition (20) is more general than Hahn's condition for the 2-transformation, but only slightly more general than his own condition for the 1-transformation (12).[9]

In §5 and §6 we saw that the step from the ordinary Fourier(–Plancherel) transformation to the 1-transformation gave a major extension to the theory, namely from the monotonically decreasing functions (or L^1 or L^2 functions) to the essentially bounded functions. Why, then, did the step from the 1-transformation to the 2-transformation only give a minor extension?

One would imagine that because of the extra x and t, respectively, in the denominators of (17), (18), (21) and (22) one could treat functions tending to infinity slower than x and t, respectively. The reason that this does not work is to be found in the inversion formulas (16) and (23). It is seen most clearly in Wiener's case where the existence a.e. of the derivatives Δ' and Γ' is used explicitly. Thus, in order for the inversion formula to make sense, Γ and Δ must be absolutely continuous, a requirement which naturally restricts the allowable fs. Therefore, in order to progress with the generalizations, the inversion formula had to be abandoned. This was done by Bochner.

12. Bochner began his work on the generalized Fourier transformation in an article written together with Hardy [1926], presenting a simplification of Wiener's proofs of the formulas (13), (14) and (15).

Whereas Hahn and Wiener had taken the first two steps in extending the Fourier transformation, Bochner took the next infinitely many steps. He did so in [1927] in his article "Darstellung reellvariabler und analytischer Funktionen durch verallgemeinerte Fourier- und Laplace-Integrale". In 1932 he included the theory in his famous textbook *Vorlesungen über Fouriersche Integrale.*[10]

In the 1927 article the theory was presented in isolation from other mathematical fields. In his textbook, on the other hand, the introduction of generalized Fourier integrals was motivated by his desire to solve the difference-differential equation

$$\sum_{\rho=0}^{r} \sum_{\sigma=0}^{s} a_{\rho\sigma} y^{(\rho)}(x + \delta_\sigma) = f(x). \tag{24}$$

Bochner first treated this equation [1932, Ch. 5] using the ordinary Fourier transformation, but it turned out that only under severe restrictions did a Fourier transformable solution exist [Satz 29, p. 96]. Thus he introduced the generalized Fourier integrals in order to weaken the conditions on the solutions y (see §17).

The difference between the 1927 article and the textbook lies not only in the motivation. Bochner [1952] explains: "In these early papers we followed a lead of Norbert Wiener in considering functions which were locally L_2-integrable instead of L_1-integrable, and this made the theory so much more complicated than the one we arrived at eventually in 'Fouriersche Integrale' that we did not emphasize them afterwards".

However, from our point of view the two approaches share their main ideas in common. Thus I shall, with a few exceptions, only discuss the theory as presented in the textbook.

13. Bochner defined the spaces F_k consisting of all functions for which

$$\int_{-\infty}^{\infty} \frac{|f(x)|}{1 + |x|^k}\, dx \tag{25}$$

exists. For a function f in F_k the k-transform $E(\alpha, k)$ can be defined by

$$E(\alpha, k) \stackrel{k}{\asymp} \frac{1}{2\pi} \int_{-\infty}^{\infty} f(x) \frac{e^{-i\alpha x} - L_k}{(-ix)^k}\, dx. \tag{26}$$

Here L_k is a polynomial multiplied by the characteristic function $\chi[-1, 1]$, which takes care of the singularity at zero; $\stackrel{k}{\asymp}$ means equality modulo an additive polynomial of degree less than $(k - 1)$. Thus, except for the complex exponential form of the integral, the 1- and 2-transformations are equivalent to Hahn's and Wiener's generalizations. Formally (or in distribution language) $d^k E(\alpha, k)/(d\alpha)^k$ is the usual Fourier transform of f, a theorem which Bochner showed under the assumption that f is in F_0 (i.e. f has the usual Fourier transform). When two $E(\alpha, k)$s differ from each other by a polynomial of degree less than $(k-1)$, then their kth derivatives are equal. Therefore it is not surprising that Bochner introduced the equivalence relation $\overset{k}{\asymp}$ in the space called T_k of all k-transforms.

He wrote symbolically

$$f(x) \sim \int e^{ix\alpha}\, d^k E(\alpha, k),^{11} \tag{27}$$

but, as pointed out before he could only obtain a real inversion formula for well behaved $E(\alpha, k)$s: namely if $E(\alpha, k)$ is k-times differentiable in the neighbourhood of ∞ and $-\infty$, the expression which will arise from k-times formal partial integration of

$$\int_{-A}^{A} e^{ix\alpha}\, d^k E(\alpha, k)$$

will tend to $f(x)$ for A tending to infinity *if the limit exists a.e.* [1932, §31]. In other words, the generalized expression (27) cannot converge to anything but f.

> Eine eigentliche Verallgemeinerung der Wienerschen Formel [i.e. formulas (15) and (23)] auf Funktionen höherer Klassen hat Verfasser [Bochner] nicht finden können. [Bochner 1927, p. 652.]

14. The symbols $d^k E(\alpha, k)$ are to a limited degree considered separately from the expression (27) by Bochner. These symbols are equivalent to distributions, namely to the Fourier transforms of the functions f in F_k. Moreover, as was pointed out by Schwartz (§1), they can be considered as the derivatives

$$\frac{d^k E(\alpha, k)}{d\alpha^k} \quad \text{in } \mathscr{D}'$$

of the continuous functions $E(\alpha, k)$. However, Bochner has not accounted for all distributions, because only locally is a distribution a derivative of a continuous function. Alternatively: the $d^k E(\alpha, k)$s only represent the Fourier transforms of ordinary *functions* in $\mathscr{S}'$. Since the Fourier transform is a bijection of $\mathscr{S}'$ onto itself, all the Fourier transforms of temperate distributions which are *not* functions are missing.

This gives an asymmetry in Bochner's theory, which he pointed out clearly in [1927, Nachtrag]. In this early work $F_0 = T_0$ because of Plancherel's theorem (recall that in 1927 Bochner worked in L^2 and not in L^1). However, as Bochner pointed out, the inclusions $T_k \subset F_k$ are not identities for $k > 0$. Since if φ is a nondifferentiable F_0-function with the $(k-1)$ transform $E(\alpha, k-1)$, then $\alpha E(\alpha, k-1)$ is in F_k but does not belong to T_k. The modern explanation is that $\alpha(d^{k-1}/d\alpha^{k-1})E(\alpha, k-1)$ is the Fourier transform of φ' (modulo a constant) which does not belong to F_k. Thus the asymmetry was recognized by Bochner, but he did not try to overcome it—or at least he did not succeed.

15. It is interesting to see that Bochner in his review [1952] of Schwartz' *Théorie des Distributions* reduced Schwartz' conceptual innovation to the establishment of this symmetry. I shall cite passages from the review, not for the purpose of discussing Schwartz' work (that will be done in Ch. 6), but to show Bochner's personal view of it, which will give an excellent account of the logical connection between his and Schwartz' work:

In Euclidean E_k we consider a general function $\varphi(x) = \varphi(x_1, \ldots, x_k)$ which is defined and infinitely differentiable everywhere and is zero outside a bounded domain $D = D_\varphi$, and, as in a previous context, we call such a function a *testing function.*[12] Next, if $F = F(x)$ is a fixed function in E_k which at first is continuously differentiable of order not less than n, then, by partial differentiation, we obtain, for $k = 1$, for the integral

$$\int_{-\infty}^{\infty} \varphi \frac{d^n F}{dx^n}\, dx \tag{28}$$

the value

$$(-1)^n \int_{-\infty}^{\infty} F \frac{d^n \varphi}{dx^n}\, dx, \tag{29}$$

and for general k we obtain for

$$\int_{E_k} \varphi \frac{\partial^{n_1 + \cdots + n_k} F}{\partial^{n_1} x_1 \cdots \partial^{n_k} x_k}\, dv_x \tag{30}$$

the value

$$(-1)^{n_1 + \cdots + n_k} \int_{E_k} F \frac{\partial^{n_1 + \cdots + n_k} \varphi}{\partial^{n_1} x_1 \cdots \partial^{n_k} x_k}\, dv_x. \tag{31}$$

Now, for the computation of the integrals (29), (31) the function F need not be differentiable and this leads to defining the symbols (28), (30) for testing functions φ (in terms of their values (29), (31)) even if the differentiation on F cannot be carried out literally. Such generalized integrals have been long in developing, and their systematic use was the very basis for the theory of generalized Fourier transforms as presented in the reviewer's book *Fouriersche Integrale*, 1932.

Bochner continued to explain that Schwartz had shown that the formal differential quotients were equivalent to certain functionals at least locally; however, it is not clear from Bochner's review that Schwartz actually used the latter and not the former as the *definition* of distributions. After this Bochner briefly sketched Schwartz' main result in the theory of Fourier series: A periodic distribution has a Fourier series that converges to the distribution in $\mathscr{D}'$. Conversely, any trigonometric series with coefficients of slow growth is a Fourier series of a distribution. The amazingly simple proof of this strong theorem underlines the genius of Schwartz' approach. Bochner, however, does not favour this point of view in his comments on Schwartz' theorem:

The author rather prides himself on that last statement, but within the given context it amounts only to stating that....

He continued:

> The second half of volume II is given over to generalized Fourier integrals, and there the analysis had always been very much subtler, *and has so remained.*

wrote Bochner and continued with a one page account of his own work in the field. He slightly altered the symbolic expression (27) to read

$$f(x) \sim \int_{-\infty}^{\infty} e^{ix\alpha} \frac{d^n g_n(\alpha)}{d\alpha^n}\, d\alpha, \qquad \text{(see note 11)}, \tag{32}$$

which is more easily compared with the theory of distributions, and commented:

> The resulting function $g_n(\alpha)$ will not be differentiable any more, but we nevertheless envisaged the relation (32) in a symbolic fashion, and these were the generalized Fourier integrals in our book cited....
>
> Now, turning again to our relation (32) we note that the author [Schwartz] goes a step further than we did and he also differentiates the function $f(x)$ itself symbolically, thus
>
> $$\frac{d^p f(x)}{dx^p} \sim \int_{-\infty}^{\infty} e^{ix\alpha}(i\alpha)^p \frac{d^n g_n(\alpha)}{d\alpha^n}\, d\alpha.$$
>
> At first sight this still leaves the two sides unsymmetric in that, seemingly, on the right side the integrand is
>
> $$(i\alpha)^p \frac{d^n g_n(\alpha)}{d\alpha^n},$$
>
> with the unbalancing factor $(i\alpha)^p$ in front. It turns out however that for a suitable $G(\alpha)$ and m this can be symbolically written as
>
> $$\frac{d^m G(\alpha)}{d\alpha^m};$$
>
> and in this way the author arrives at a symmetric Fourier transformational reciprocity between symbols
>
> $$\frac{d^p F(x)}{dx^p}, \qquad \frac{d^m G(\alpha)}{d\alpha^m}$$
>
> as it were, the functions $F(x)$, $G(x)$ being arbitrary continuous functions in $(-\infty, \infty)$ which are $0(|x|^q)$ and $0(|\alpha|^n)$ at infinity, the indices p, m, q, n being unrestricted.
>
> ....
>
> We note however that the resulting self-inversiveness of the class of distributions [i.e. the self-inversiveness of the Fourier transformation], interesting as it is, is only the "dual" to the self-inversiveness of the semi-testing functions themselves, which latter self-inversiveness is a rather obvious phenomenon and, for instance, cannot compare in subtlety to the self-inversiveness of the Plancherel transforms, say, where a natural norm is preserved as well.

Thus Bochner did not grant Schwartz much recognition for the conceptual development of the theory of distributions.

16. Moreover, he pointed out that some of the technical tools were not original with Schwartz either:

> A dominant analytical tool in the work is a certain "smoothing" process (in French "régularisation") which is used both to localize pieces of a spectrum or of a functional, and to approximate to a distribution by a function. An an analytical tool it is older than sometimes realized and it has been constantly used by us both for generalized Fourier integrals and almost periodic functions; and the closely related "partition of unity", so-called, which is gaining in importance in the cohomology theory of differential forms was introduced for the first time for just such a purpose in our note: "Remark on the theorem of Green", *Duke Math. J.*, **3** (1937), pp. 333–338.[13] And as regards the novelty of introducing "distributions" which are more general than Stieltjes integrals, say, we think that the credit for it ought to be assigned to Riemann.

Thus when Bochner could not take the credit himself, he bestowed it on one of the mathematical giants of the past (and not on Sobolev for example). However, as I pointed out in Ch. 2, note 18, Riemann does not deserve this honour. Bochner concluded:

> We have recounted all this with a view to suggesting that it would not be easy to decide what the general innovations in the present work are, analytical and even conceptual, and that it is in order to appraise the value of the book by its specific results, such as we have extracted above; and of such let the author produce many more, by all means.

Let me end this account of Bochner's review by mentioning a few figures; out of a total of 12 pages, $6\frac{1}{2}$ pages are devoted to Schwartz' theory of distributions exclusively, $1\frac{1}{2}$ to Bochner's own theory and the rest to a comparison of the two theories.

17. It has become quite clear from the review that there is logically a strong link between Bochner's generalized Fourier integrals and the theory of distributions. Moreover, as already pointed out (§ 1), Schwartz underscored the similarity himself. What then are the differences between the two theories? How close did Bochner come to a theory of distributions?

Distributions are not in fact automatically introduced by the symbolic expression (27) as it stands. The question remains whether the symbol $d^kE(\alpha, k)$ is detached from the integral expression, obtaining a meaning by itself. To answer this question one must also answer the related question: Did Bochner introduce any algebraic, analytic or topological notions for the $d^kE(\alpha, k)$s which allowed him actually to *operate* with them? The answer to both questions is that each occurred, but only to a very limited degree: namely to the degree which allowed Bochner to treat the difference-differential equation (24) by means of the generalized Fourier transformation. If the usual Fourier transform of y in (24) is $\varphi(\alpha)$ and of f is $E(\alpha)$, then the equation (24) yields the following equation between the Fourier transforms:

$$G(\alpha)\varphi(\alpha) = E(\alpha) \tag{33}$$

where

$$G(\alpha) = \sum_{\rho=0}^{r} \sum_{\sigma=0}^{s} a_{\rho\sigma}(i\alpha)^{\rho} e^{i\delta\sigma\alpha}. \tag{34}$$

The formally equivalent equation for the k-transform would be

$$G(\alpha)\, d^k\Phi(\alpha, k) = d^k E(\alpha, k), \tag{35}$$

where $\Phi(\alpha, k)$ is the k-transform of y and $E(\alpha, k)$ is the k-transform of f. In order to give (35) meaning, Bochner had to define what he meant by a function multiplied by a symbol $d^k\Phi(\alpha, k)$. He specified that

$$d^k\varphi(\alpha) = \chi(\alpha)\, d^k\psi(\alpha) \tag{36}$$

should mean:

$$\varphi(\alpha) \stackrel{k}{\asymp} \chi(\alpha)\psi(\alpha) + \sum_{\kappa=1}^{k} (-1)^{\kappa} \binom{k}{\kappa} \int_{(\kappa)}^{\alpha} \chi^{(\kappa)}\psi(\alpha)\, d\alpha, \tag{37}$$

where the integral is the κ-times iterated integral. ((37) can be obtained from (36) by formal partial integration.)

With this definition (35) actually follows from (24) when y and f lie in F_k. Bochner was able to prove [Satz 45] that (35) always had a solution Φ in T_k for k large enough, i.e. that (24) had a solution in $\bigcup_{k=0}^{\infty} F_k$.

Thus Bochner did separate the expressions $d^kE(\alpha, k)$ from the integral (27) and he did introduce one "operation", namely, multiplication by a function. However he only used the $d^kE(\alpha, k)$s in connection with the integral (27) or in equations like (35) and only introduced this one operation.[14] Thus one must conclude that $d^kE(\alpha, k)$ only had a meaning in connection with the symbolic expression (27) (and sometimes (35)). The entities which assumed the role of the Fourier transforms were the functions $E(\alpha, k)$ and not their kth derivatives as the review of Schwartz' book might suggest. Bochner did not operate with *one* generalized Fourier transformation and transformed "distributions" of different orders of irregularity, but rather he operated with different transformations—the k-transformations—for which the transforms were ordinary functions $E(\alpha, k)$. Therefore, even though Bochner considered symbols which were equivalent to distributions in connection with the Fourier integrals, one cannot say that he possessed a theory of distributions.

18. Bochner never applied his "distributions" outside of the theory of Fourier transformations. For instance, in his 1946 article on the theory of differential equations, he gave a method of generalization which had nothing to do with the generalized Fourier transform [Bochner 1946] and in his 1932 textbook he considered only ordinary solutions. Had he applied his symbols $d^kE(\alpha, k)$ to describe generalized solutions it would have required that they not be tied to Fourier integrals and that several operations be defined for these objects. If, for instance, Bochner had treated generalized solutions of the form $d^kE(\alpha, k)$ in 1932, he would have been forced to define the Fourier integral for these objects. This in turn would have given him the symmetry which he could not find for his own version of the generalized Fourier integral.

19. We have discussed the logical connection between Bochner's and Schwartz' theories. The historical connection was explained by Schwartz in his autobiography [1974]. While explaining his discovery of distribution theory in 1945, he wrote:

> J'ignorais alors les travaux de S. Bochner....

So Schwartz was not inspired by Bochner's work when he created the distribution theory.

In this same note Schwartz wrote about Bochner's work:

> S. Bochner a introduit, dans son livre sur l'intégrale de Fourier en 1932, sur la droite réelle, des "dérivées formelles de produits de polynômes par des fonctions de L^2"; ce sont mes future distributions tempérées. Il en fait la transformation de Fourier.

Schwartz seems to confuse things here. The products of polynomials and L^2 functions are Bochner's F_n^2 functions (as pointed out, Bochner usually used polynomials multiplied by L^1 functions, i.e. F_n functions, but he has a few remarks about F_n^2 as well) and the Fourier transformation is introduced for such *functions*. On the other hand, Bochner's formal derivatives are the Fourier transforms of functions of $F_n^{(2)}$. He did not define generalized Fourier transforms of such formal derivatives. Bochner even admitted this in his review [1952] (see §15) of Schwartz' *Théorie des Distributions*. Thus Schwartz overestimated what Bochner had actually done. On the other hand, he is quite explicit about other things Bochner did not do:

> il ne semble pas lui-même y attacher beaucoup d'importance. Le support n'y est pas, ni aucune topologie, ni les conditions de transformation du produit de convolution, et finalement la distribution de Dirac δ n'y est pas nommé.

20. The problem of extending the Fourier transformation to functions of polynominal growth, with which Hahn, Wiener and Bochner had struggled, was also attacked in quite a different way by the Swedish mathematician T. Carleman. His solution to the problem was published in his book, *L'Intégral de Fourier et Questions qui s'y Rattachent* [1944, spec. Ch. II], which is a summary of a course he presented at the Mittag-Leffler Institute in 1935.

Carleman first remarked that for an $L^1(\mathbb{R})$ function f the Fourier transform g could be split into two parts, namely:

$$g(z) = \frac{1}{\sqrt{2\pi}} \int_{-\infty}^{\infty} e^{-izy} f(y)\, dy = g_1(z) - g_2(z), \tag{38}$$

where

$$g_1(z) = \frac{1}{\sqrt{2\pi}} \int_{-\infty}^{0} e^{-izy} f(y)\, dy$$

and

$$g_2(z) = -\frac{1}{\sqrt{2\pi}} \int_0^{\infty} e^{-izy} f(y)\, dy. \tag{38a}$$

He took z to be a complex variable and saw that even if f were not L^1 but only satisfied the condition

$$\int_0^x |f(y)|\, dy = 0(|x|^{\kappa}) \quad \text{for a natural number } \kappa, \tag{39}$$

then g_1 was an analytic function for $I_m(z) > 0$ and g_2 was an analytic function for $I_m(z) < 0$.

> Cela posé nous allons, dans ce cas, définir la transformée de Fourier généralisée de de $f(x)$ comme la paire des fonctions analytiques $g_1(z)$ et $g_2(z)$.

Since

$$g_1(\alpha + i\beta) - g_2(\alpha - i\beta) \quad \text{for } \beta > 0 \tag{40}$$

is the ordinary Fourier transform of the function $e^{-\beta|x|}f(x)$, Carleman could recover f from g_1 and g_2 by taking the inverse transform of (40) and multiplying by $e^{\beta|x|}$. However, he was not satisfied with the asymmetry between the generalized Fourier transformation and its inverse. Therefore he constructed a Fourier transformation operating between spaces of function-pairs.

In order to show that this procedure generalized the ordinary Fourier transformation, he first proved that a function f satisfying (39) in a unique way gave rise to a pair of functions. More precisely, he showed that for such a function f there existed analytic functions $f_1(z)$ and $f_2(z)$ regular for Im $z > 0$ and Im $z < 0$, respectively, such that

$$\lim_{y \to 0} \int_{x'}^{x''} (f_1(x + iy) - f_2(x - iy))\, dx = \int_{x'}^{x''} f(x)\, dx \tag{41}$$

uniformly in every domain

$$a \leq x' \leq x'' \leq b$$

for a, b any fixed real numbers.

In other words, f can be represented as the jump from f_2 to f_1 on the real axis. However not all such jumps represent functions. Therefore function pairs such as those considered by Carleman constitute a generalization of functions satisfying (39) (tempered functions).

For function pairs satisfying the growth conditions

$$|f_1(re^{i\theta})| < A(\theta_0)\left(r^{\alpha} + \frac{1}{r^{\beta}}\right) \quad \text{for } \theta_0 < \theta < \pi - \theta_0,$$

$$|f_2(re^{i\theta})| < A(\theta_0)\left(r^{\alpha} + \frac{1}{r^{\beta}}\right) \quad \text{for } -\pi + \theta_0 < \theta < -\theta_0,\ \alpha, \beta \geq 0, \tag{42}$$

Carleman introduced the Fourier transformation in the following way: first two functions $G(z)$, $H(z)$ are defined by

$$G(z) = \frac{1}{\sqrt{2\pi}} \int_L e^{-izy} f_1(y)\, dy, \tag{43}$$

$$H(z) = \frac{1}{\sqrt{2\pi}} \int_{L'} e^{-izy} f_2(y)\, dy, \tag{44}$$

where L and L' are rays emanating from the origin in the upper and lower half-plane, respectively; rotating L and L' around the origin, the integrals G and H can be defined over the whole complex plane except for the positive and negative real axes.[15]

> Cela posé nous allons introduire la paire de fonctions $g_1(z)$ et $g_2(z)$ qui représente la transformée de Fourier généralisée de $f_1(z)$ et $f_2(z)$ par les relations:

$$\begin{aligned} g_1(z) &= H(z) - G(z) \quad \text{for } \operatorname{Im}(z) > 0, \\ g_2(z) &= H(z) - G(z) \quad \text{for } \operatorname{Im}(z) < 0. \end{aligned} \tag{45}$$

The Fourier transformed couple again satisfies the inequality (42).[16] Carleman used the letter S to denote the linear transformation carrying the f-pair into the g-pair. Moreover, he introduced another linear operator T taking $(g_1(z), g_2(z))$ into $(\overline{g_1(\bar{z})}, \overline{g_2(\bar{z})})$. Then he stated the generalized inversion formula

$$TSTS(f) = f. \tag{46}$$

Thus, in contrast to Bochner, Carleman obtained a beautiful symmetric inversion formula for the generalized Fourier integrals.

21. As in Bochner's case the question arises: How close was Carleman to a theory of distributions? I shall first examine the historical facts and afterwards investigate the logical mathematical relations between Carleman's work and distribution theory.

From his statement of the representability of a "tempered function" (39) by a function pair there can be no doubt that Carleman knew that his function pairs represented a generalization of the function concept. Carleman's function pairs existed much more independently of their use in Fourier analysis than Bochner's symbols $d^k E(\alpha, k)$. In particular, there was no question about how to operate with function pairs since the usual rules for complex numbers could be applied to each component separately. Nevertheless, Carleman seems to have attached little importance to his function pairs.

He applied the generalized Fourier transformation to solve an integral equation of the first kind [1944, Note II], but he did not apply his "generalized functions" outside of the field of Fourier transformations.

Carleman and Schwartz met in 1947 at the Colloque International de Analyse Harmonique in Nancy where they both gave papers on their generalization of the Fourier integrals to function pairs and tempered distributions respectively. However, neither of them later tried to connect the two theories. Schwartz immediately thought that their works were in some sense isomorphic, but he made no attempt to prove this.[17] [Schwartz 1978, Interview.]

22. The connection between the two theories was to some degree established in the theory of hyperfunctions which was developed in the 1950s and 1960s. In this theory the function concept is generalized precisely in the way that Carleman did, namely, by considering pairs of analytic functions or functions analytic in the upper and lower half-plane (see Appendix, §5). The notion of a hyperfunction is a proper generalization of the concept of a distribution on the real axis. It has been shown by Bremermann [1965, p. 50] that for every distribution $T \in \mathscr{D}'(\mathbb{R})$ there exists a function analytic except on the support of T for which

$$f(x + i\varepsilon) - f(x - i\varepsilon) \underset{\varepsilon \to 0}{\to} T \quad \text{in } \mathscr{D}'. \tag{47}$$

(Less general representation theorems were proved by Tillmann [1961b] and Bremermann and Durand [1961]). This generalizes Carleman's representation theorem (41). On the other hand, there exist analytic functions which do not represent distributions in the above way. One example is e^{-1/z^2} [Bremermann 1965, p. 70]. Thus the generalized functions used by Carleman are even more general than Schwartz' distributions.[18]

However, the subclass of function pairs for which Carleman defined the Fourier transformation is subject to condition (42). The question then arises: How large a class of hyperfunctions satisfies this condition? It is easy to show that all tempered distributions belong to Carleman's class of function pairs. Tillmann in [1961b] gave the growth condition:

$$|f(z)| \leq M_0(1 + |z|^2)^m \cdot |\mathrm{Im}\, z|^{-n} \tag{48}$$

characteristic of the holomorphic functions corresponding to tempered distributions, and it is easily checked that (48) implies (42). On the other hand, the function pair

$$f_1(z) = 0 \quad \text{for } \mathrm{Im}\, z > 0,$$

$$f_2(z) = \exp\left[iz + (\log z)^2 \frac{z}{z + i}\right] \quad \text{for } \mathrm{Im}\, z < 0. \tag{49}$$

gives an example of a pair satisfying Carleman's conditions, but not equivalent to a tempered distribution.[19,20] Thus Carleman introduced the Fourier transformation in a space which is effectively larger than the tempered distributions.

23. At the Colloque International 1947, where both Carleman and Schwartz gave their generalizations of the Fourier integral, the Swede Arne Beurling gave a third method, which however is closely related to Carleman's [Beurling 1947]. For any measurable function f satisfying

$$\int_{-\infty}^{\infty} |f(x)| e^{-\sigma|x|}\, dx < \infty \tag{50}$$

for all $\sigma > 0$ he defined the "transformée harmonique"

$$U_f(\sigma, t) = \int_{-\infty}^{\infty} f(x) e^{-itx-\sigma|x|}\, dx \qquad \sigma > 0. \tag{51}$$

He remarked that if F denotes the Laplace transform:

$$\begin{cases} \displaystyle\int_0^{\infty} f(x) e^{-x(\sigma+it)}\, dx = F_1(\sigma + it) & \sigma > 0 \\ \displaystyle -\int_{-\infty}^{0} f(x) e^{-x(\sigma+it)}\, dx = F_2(\sigma + it) & \sigma < 0. \end{cases} \tag{52}$$

then

$$U_f(\sigma, t) = F_1(\sigma + it) - F_2(-\sigma + it). \tag{53}$$

Since

$$F_1(\sigma + it) = -g_2(t - i\sigma) \tag{54}$$

and

$$F_2(\sigma + it) = g_1(t - i\sigma), \tag{55}$$

where (g_1, g_2) is Carleman's function pair defined by (38a), one obtains

$$U_f(\sigma, t) = g_1(t + i\sigma) - g_2(t - i\sigma). \tag{56}$$

Therefore, if f satisfies Carleman's condition (39), the boundary value of $U_f(\sigma, t)$ for $\sigma \to 0$ represents the same generalized function as the jump between Carleman's holomorphic functions g_1 and g_2. Beurling, however, did not develop a symmetric theory as Carleman did, and thus did not come as close to a theory of generalized functions.

Chapter 4
Early Generalized Functions

Part 1. Fundamental Solutions. Green's Function

1. In *Théorie des Distributions* a fundamental solution with respect to a point a and a differential operator L is defined as any distribution E_a satisfying

$$L(E_a) = \delta_a = \text{the mass 1 at the point } a. \tag{1}$$

The importance of the fundamental solution rests on the fact that once it has been found, the solution to the equation

$$L'(\varphi) = \psi \qquad (L' \text{ being the adjoint of } L) \tag{2}$$

for any $\psi \in \mathscr{D}$ on the right is given by the simple expression

$$\varphi(a) = E_a(\psi). \tag{3}$$

If L has constant coefficients, the fundamental solution with respect to a can be found from the fundamental solution $E = E_0$ with respect to 0 by a simple translation. In that case the solution $T \in \mathscr{D}'$ to

$$L(T) = B, \tag{4}$$

with $B \in \mathscr{E}'$ is given by

$$T = E * B. \tag{5}$$

This is obvious from the identity

$$L(E * B) = L(E) * B = \delta * B = B. \tag{6}$$

The modern theory of fundamental solutions is thus so dependent of the notion of a δ-distribution that it may be astonishing to realize that its roots extend as far back as 1828. At that time George Green (1793–1841) introduced the so-called Green's function as a means for studying boundary value problems in electrostatics. When boundary conditions are imposed on the fundamental solutions, they are usually called Green's functions after their inventor. However, the meaning of the terms Green's function and fundamental (or elementary) solution varied from author to author. I shall simply apply the terms as they were used by the different authors under consideration, as long as no confusion will result from such a procedure.

Prior to 1950, however, the fundamental solution was not given by an equation similar to (1) but was defined as a solution with certain singularities. Why then is the development of fundamental solutions of interest in the prehistory of the theory of distributions? There are two reasons for this. First, the δ-function still played various roles in connection with the fundamental solution; it was actually as part of the definition of the fundamental solution of the wave equation that the δ-function was introduced for the first time. Secondly, Hadamard's "partie finie"—another essential notion in the theory of distributions—was invented to avoid certain divergent integrals which appeared when the fundamental solutions were used. The first sections (§2–10) of this part deal with those aspects of the development which are of interest in the history of the δ-function and the last sections (§11–15) are devoted to the partie finie.

2. How can a fundamental solution be defined without the use of the δ-function? This question is beautifully answered in Courant and Hilbert's *Methoden der Mathematischen Physik, I* [1924]. Courant first treated the simple case of a second-order ordinary differential equation [p. 274]

$$L(y) = py'' + p'y' + qy = \varphi(x). \tag{7}$$

In the interval $\Omega = [x_0, x_1]$ a solution is sought which satisfies homogenous boundary conditions $f(x_0) = f(x_1) = 0$. Courant started with a physical *analysis* of the problem, interpreting (7) as the equation of equilibrium of a string subject to the external force $\varphi(x)$:

> Machen wir nun einen Grenzübergang von der kontinuierlich verteilten Kraft $\varphi(x)$ zu einer "Einzelkraft", d.h. einer nur in einem einzigen Punkte $x = \xi$ mit der Identität 1... angreifenden Kraft, und ist $K(x, \xi)$ die Elongation der Saite unter dem Einfluss dieser Einzelkraft, wobei stets die der Saite auferlegten Randbedingungen gewahrt bleiben sollen, so wird man die Wirkung die kontinuierlich verteilten Kraft $\varphi(x)$ als Superposition der Wirkungen kontinuierlich verteilten Einzelkräfte auffassen können, deren Dichte an der Stelle x' gleich $\varphi(x')$ ist; man kann also erwarten dass die gesuchte Lösung der Form
>
> $$f(x) = \int_{x_0}^{x_1} K(x, x')\varphi(x')\,dx' \tag{8}$$
>
> erscheint.[1,2]

This imprecise introduction of the Green's function $K(x, x')$ corresponds to equation (1).

In order to find a mathematical characterization of K, Courant introduced an approximation to the unit force $\delta(x - x')$, that is a sequence φ_ε satisfying

$$\varphi_\varepsilon(x) = 0 \quad \text{for } |x - x'| > \varepsilon, \qquad (9)$$
$$\int_{x'-\varepsilon}^{x'+\varepsilon} \varphi_\varepsilon(x) = 1.$$

The corresponding elongation $K_\varepsilon(x, x')$ satisfies

$$L_x K_\varepsilon(x, x') = \varphi_\varepsilon(x). \qquad (10)$$

By integration of (10) Courant found.

$$\int_{x'-\varepsilon}^{x'+\varepsilon} \left(p \frac{d^2K_\varepsilon}{dx^2} + p' \frac{dK_\varepsilon}{dx} + qK_\varepsilon\right) dx = 1$$

and for $\varepsilon \to 0$ he obtained (assuming K to be continuous)

$$\lim_{\varepsilon \to 0} \frac{dK_\varepsilon(x, x')}{dx}\bigg|_{x=x'-\varepsilon}^{x=x'+\varepsilon} = \frac{1}{p(x')}. \qquad (11)$$

This led to the following mathematically satisfactory definition of the Green's function K:

(a) $K(x, x')$ is a continuous function of x for constant x';
(b) K'_x and K''_{xx} are continuous in Ω except for the point $x = x'$ where K'_x makes a jump given by

$$\frac{dK(x, x')}{dx}\bigg|_{x=x'-0}^{x=x'+0} = \frac{1}{p(x)};$$

(c) K satisfies $L_x(K(x, x')) = 0$ in $\Omega\backslash x'$.

When (a) is not fulfilled but (b) and (c) are satisfied Courant called K a "Grundlösung". I shall not make this distinction in what follows. Having found the definition of K by this physical analysis, Courant proved *synthetically* that f given by (8) actually satisfied the equation (7). If, moreover, $K(x, x')$ has homogeneous boundary values, it is clear that so does f. For boundary values $u = u_0 \neq 0$ on $\partial\Omega$ Courant easily reduced the boundary value problem to the homogeneous case by extending u_0 to a sufficiently smooth function in all of Ω (if possible) and then solving the equation for $u - u_0$.

Courant and Hilbert provided the same analysis for higher-order ordinary differential equations and for the Laplace equation and found the appropriate singularities which could be used in a definition of the Green's function along lines similar to (a), (b), and (c) above. Thus he obtained rigorous definitions for the Green's functions for several equations. However, one looks in vain for a general definition of a Green's function or "Grundlösung". The only property characterizing all the Green's functions in [Courant–Hilbert 1924, 1937] is the physical description of them as states resulting from external unit forces. The theory of distributions supplied this want of a general definition of Green's function and of the fundamental solution, namely definition (1). With this theory the definition was brought into accordance with the intuitive idea which Courant had used in his physical analysis of the problem. Moreover, with distribution theory the synthetic proof for (8) could be avoided since reversing the analysis as in (6) gave a rigorous proof. Thus the role played by the theory of distributions in connection with the fundamental solutions was not primarily that of rigorization. Rather it was one of unification and of bringing mathematical and intuitive physical arguments into accord with each other.[3]

3. Having seen how fundamental solutions can be defined without the use of the δ-function, we will return to the beginning of the story, to Green's (1793–1841) revolutionary *An Essay on the Application of Mathematical Analysis to the Theory of Electricity and Magnetism* [1828]. This was the first work on electrostatics in which mathematical analysis played a decisive role. Green's main result gives a formula for determining the electric potential V in a domain in a vacuum bounded by conductors with given potentials. Green showed that the corresponding mathematical problem asks for a solution to $\Delta V = 0$ in a domain Ω satisfying given boundary conditions on $\partial\Omega$. To solve this problem Green first proved Green's theorem

$$\int_\Omega U \Delta V \, d\bar{x} + \int_{\partial\Omega} U \frac{dV}{dn} \, d\sigma = \int_\Omega V \Delta U \, d\bar{x} + \int_{\partial\Omega} V \frac{dU}{dn} \, d\sigma. \tag{12}$$

It was proved for U and V having bounded derivatives (obviously this is not restrictive enough). However his trick was to use it for a solution $U(\bar{x}, \bar{x}')$ to $\Delta_x U = 0$ regular in $\Omega \setminus \{\bar{x}'\}$ and having, in $\bar{x}'$, a singularity of the form $1/(\bar{x} - \bar{x}')$. Written in the more precise way:

(a) $U(\bar{x}, \bar{x}') \in C_x^2(\Omega \setminus \{\bar{x}'\})$;

(b) $U(\bar{x}, \bar{x}') = [1/(|\bar{x} - \bar{x}'|)] + \gamma(\bar{x}, \bar{x}')$ where $\gamma \in C_x^2(\Omega)$; (13)

(c) $\Delta_x U(\bar{x}, \bar{x}') = 0$ for $\bar{x} \in \Omega \setminus \{\bar{x}'\}$.

this is (except for a factor 4π) the characterization Courant later gave for the Green's function for the Laplace equation. Green had no special name for U. It was Riemann who named it Green's function.

$U(\bar{x}, \bar{x}')$ satisfies

$$\Delta U(\bar{x}, \bar{x}') = -4\pi\delta(\bar{x} - \bar{x}'). \tag{14}$$

Green could not use (12) directly on this singular function. Thus he applied the trick, which has now become standard, of removing a small ball about $\bar{x}'$. The singularity in U contributed to the surface integral over the small ball with a quantity, having the finite limit $4\pi V(\bar{x}')$ for the radius of the ball tending to zero. Green had then obtained:

$$\int_\Omega U\Delta V\,d\bar{x} + \int_{\partial\Omega} U\frac{dV}{dn}\,d\sigma = \int_\Omega V\Delta U\,d\bar{x} + \int_{\partial\Omega} V\frac{dU}{dn}\,d\sigma - 4\pi V(\bar{x}'). \quad (15)$$

Here the integral over $V\Delta U$ is to be understood as a limit of the integrals where a small ball about $\bar{x}'$ is removed, i.e., at the singularity $\bar{x}'$, ΔU is to be taken equal to zero.[4]

Green required that the function U, in addition to conditions (a)–(c) should satisfy the homogeneous boundary conditions

(d) $$U(\bar{x}, \bar{x}') = 0 \quad \text{for } \bar{x} \in \partial U.$$

Thus he obtained

$$V(\bar{x}') = \frac{1}{4\pi}\int_{\partial\Omega} \bar{V}\frac{dU}{dn}\,d\sigma, \quad (16)$$

where $\bar{V}$ is the given boundary value of V on $\partial\Omega$. This would solve Green's boundary value problem if a Green's function U with the properties (a)–(d) could be found. Green took the existence of such a function for granted since he knew that physically it described the electrical potential from a point charge at $\bar{x}'$ when $\partial\Omega$ is thought of as a grounded metal surface ((a)–(c) is (1) in disguise). Existence problems had not yet become central in mathematics. Later on, however, one of the main problems in the theory of partial differential equations was to prove the existence of Green's functions for different equations and different boundary surfaces.

4. Green's essay remained relatively unknown until it was published in the *Journal für die reine und angewandte Mathematik* in 1850 and 1854, and it took additional time before the notion of a Green's function was generalized to other elliptic equations. In [1877] Carl Neumann initiated the study of the Laplace equation in the plane. He found that the two-dimensional equivalent of Green's function was described not by a singularity of the form $1/(|\bar{x} - \bar{x}'|)$ but by a singularity of the form

$$\log\frac{1}{|\bar{x} - \bar{x}'|}. \quad (17)$$

This result was in turn generalized by E. Picard and A. Sommerfeld (1891, 1900) to the equation

$$\Delta u + C(x, y)u = 0 \quad (18)$$

and by Hilbert, Hedrick and Hadamard (1901) to

$$\Delta u + A(x, y)\frac{\partial u}{\partial x} + B(x, y)\frac{\partial u}{\partial y} + C(x, y)u = 0 \quad (19)$$

(cf. [Hadamard 1932], pp. 99–100]). They found that a solution of the form

$$U(\bar{x}, \bar{x}')\log\frac{1}{|\bar{x} - \bar{x}'|} + W(\bar{x}, \bar{x}') \qquad (U, W \text{ regular}) \quad (20)$$

was an appropriate Green's function in this case. Also, for the more general equations above, the importance of the Green's function is due to the fact that if substituted in a fundamental formula similar to Green's formula (12), it will give the solution to the corresponding boundary value problem. Such a generalization of Green's formula was first given by P. Du Bois-Reymond (1889) and by J. G. Darboux (1887–1896) for a second-order linear partial differential equation in two independent variables (see [Kline 1972, p. 695].

5. Riemann [1858/59] had applied a similar technique to solve an initial value problem for the hyperbolic equation

$$L(u) = \frac{\partial^2 u}{\partial x \partial y} - m\left(\frac{\partial u}{\partial x} + \frac{\partial u}{\partial y}\right) = 0 \tag{21}$$

with initial values of u and its normal derivative given on a noncharacteristic curve C. He started by proving the fundamental formula

$$\int_\Omega (vL(u) - uM(v))\, d\bar{x} = -\int_{\partial\Omega}\left[v\left(\frac{\partial u}{\partial x} - mu\right) dx + u\left(\frac{\partial v}{\partial y} + mv\right) dy\right], \tag{22}$$

where M is the adjoint of the operator L.

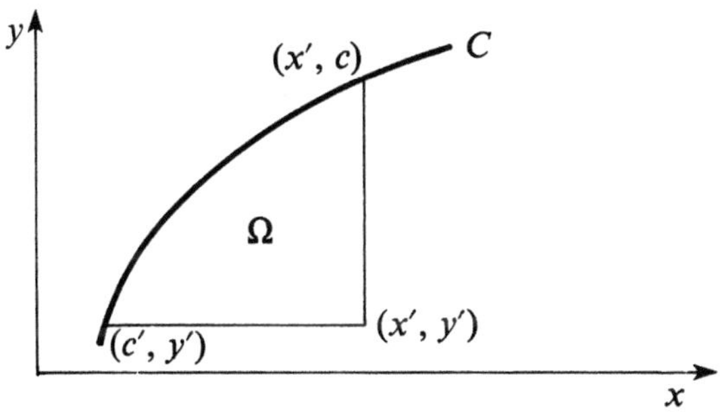

Figure 1

Taking for Ω the domain shown in Figure 1 and taking for v a function $v(x, y, x', y')$ satisfying

$$\begin{aligned}
&(1) \qquad M(v) = 0,\\
&(2) \qquad \frac{\partial v}{\partial y} + mv = 0, \quad \text{for } x = x',\\
&(3) \qquad \frac{\partial v}{\partial x} + mv = 0, \quad \text{for } y = y',\\
&(4) \qquad v(x', y') = 1,
\end{aligned} \tag{23}$$

he arrived at the expression

$$u(x', y') = (v \cdot u)(x', c) + \int_{(x', c)}^{(c', y')} \left[v\left(\frac{\partial u}{\partial x} - mu\right) dx + u\left(\frac{\partial v}{\partial y} + mv\right) dy \right] \quad (24)$$

for the desired solution at (x', y').

Although the function v, later called the Riemann function, played the same part in the integration of (21) as Green's function did in the integration of Laplace's equation, there is a profound difference between the two functions. Riemann's function has no singularity at the point $x = x'$ as Green's function has. For this reason it cannot satisfy equation (1). Hadamard, however, discovered [1932, pp. 99–101] the relationship between the two (see §8).

6. Later in the century other auxiliary functions were used by Gustav R. Kirchhoff (1824–1887) and Vito Volterra (1860–1940) in the treatment of Huygens' principle for the wave equation

$$\frac{\partial^2 u}{\partial t^2} = a^2 \Delta u \quad (25)$$

in 3 and 2 space-coordinates, respectively.

Kirchhoff [1882, §2 and 1891] started by using the original Green's theorem (12) to $u(\bar{x}, t)$, $v(\bar{x}, t)$ for fixed t and for $\bar{x}$ in a domain Ω

$$\int_{\partial\Omega} \left(u \frac{\partial v}{\partial n} - v \frac{\partial u}{\partial n} \right) d\sigma = \int_{\Omega} (v\Delta u - u\Delta v)\, d\bar{x}. \quad (26)$$

For u, v satisfying (25) he attained for the right-hand side of (26)

$$\frac{1}{a^2} \int_{\Omega} \left(v \frac{\partial^2 u}{\partial t^2} - u \frac{\partial^2 v}{\partial t^2} \right) d\bar{x} \quad \text{or} \quad \frac{1}{a^2} \frac{\partial}{\partial t} \int_{\Omega} \left(v \frac{\partial u}{\partial t} - u \frac{\partial v}{\partial t} \right) d\bar{x}$$

which after integration from $-t'$ to t'' gave:

$$\int_{-t'}^{t''} dt \int_{\partial\Omega} \left(u \frac{\partial v}{\partial n} - v \frac{\partial u}{\partial n} \right) d\sigma = \frac{1}{a^2} \left[\int_{\Omega} \left(v \frac{\partial u}{\partial t} - u \frac{\partial v}{\partial t} \right) d\bar{x} \right]_{-t'}^{t''}. \quad (27)$$

Kirchhoff then specified his auxiliary function v in the following highly interesting way:

$$v = \frac{F(|\bar{x} - \bar{x}'| + at)}{|\bar{x} - \bar{x}'|}. \quad (28)$$

Here F is a function

> die für jeden endlichen, positiven order negativen, Werth ihres Arguments verschwindet, nie negativ ist und der Bedingung genügt, dass
>
> $$\int F(\zeta)\, d\zeta = 1 \quad (29)$$
>
> wenn die Integration von einem endlichen negativen bis zu einem endlichen positiven Werthe von ζ ausgedehnt wird.

This is the first mathematical definition of the δ-function (see Part 2 of this chapter for that part of the history of the δ-function which does not relate to fundamental solutions). Thus v can be written:

$$v = \frac{\delta(|\bar{x} - \bar{x}'| + at)}{|\bar{x} - \bar{x}'|}. \tag{30}$$

This distribution is in fact the fundamental solution to the wave equation

$$\Delta_x v - \frac{1}{a^2}\frac{\partial^2 v}{\partial t^2} = -4\pi\sqrt{1 + a^2}\,\delta(\bar{x} - \bar{x}')\delta(at). \tag{31}$$

It is remarkable that this very singular "function" having δ singularities on the negative part of the light cone $|\bar{x} - \bar{x}'| + at = 0$ was already introduced by Kirchhoff.[5]

7. Kirchhoff's subsequent argument shows his reason for introducing the Green's function (28) and illustrates the way in which he manipulated the δ-function: He applied (27) to the domain $\Omega\backslash S$ where S is a small ball with center $\bar{x}'$ and radius R. Taking t' so large that $|\bar{x}' - \bar{x}| - at' < 0$ for all $x \in \Omega$

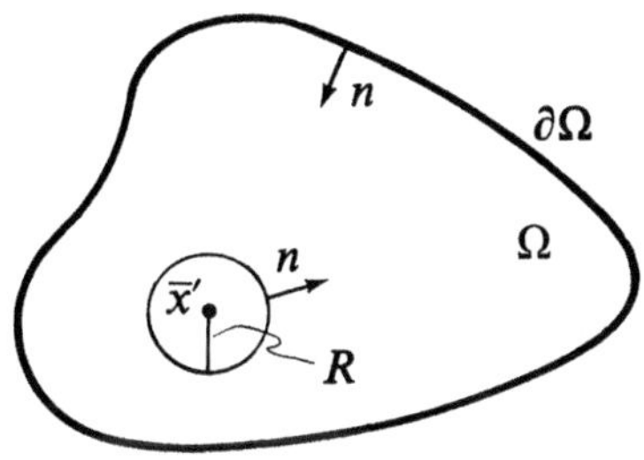

Figure 2

the right-hand side of (27) vanishes (because v vanishes) and the following equation remains.

$$\int_{-t'}^{t''} dt \int_{\partial\Omega}\left(\frac{u\partial v}{\partial n} - v\frac{\partial u}{\partial n}\right)d\sigma + \int_{-t'}^{t''} dt \int_{\partial S}\left(u\frac{\partial v}{\partial n} - v\frac{\partial u}{\partial n}\right)d\sigma = 0. \tag{32}$$

In the last integral Kirchhoff computed

$$\frac{\partial v}{\partial n} = \frac{\partial v}{\partial \bar{x}} = -\frac{1}{R^2}F(at) + o(R^2); \qquad v = o(R^2), \tag{33}$$

hence

$$\lim_{R\to 0}\int_{\partial S}\left(u\frac{\partial v}{\partial n} - v\frac{\partial u}{\partial n}\right)d\sigma = -4\pi u(\bar{x}')F(at). \tag{34}$$

From the characterization (29) of the δ-function F Kirchhoff found

$$\int_{-t'}^{t''} F(at)\,dt = \frac{1}{a}, \tag{35}$$

so that the last integral in (32) has the value

$$-\frac{4\pi}{a}\, v(\bar{x}', 0). \tag{36}$$

In the first integral of (32) Kirchhoff interchanged the order of integration. Since

$$\frac{\partial v}{\partial n} = \frac{\partial \dfrac{F(|\bar{x}-\bar{x}'|+at)}{|\bar{x}-\bar{x}'|}}{\partial n} = \frac{\partial \dfrac{1}{|\bar{x}-\bar{x}'|}}{\partial n} F(|\bar{x}-\bar{x}'|+at)$$
$$+ \frac{1}{|\bar{x}-\bar{x}'|}\frac{\partial|\bar{x}-\bar{x}'|}{\partial n}\frac{1}{a}\frac{\partial F(|\bar{x}-\bar{x}'|+at)}{\partial t}, \tag{37}$$

he obtained

$$a\int_{-t'}^{t''} u\frac{\partial v}{\partial n}\,dt = \frac{\partial \dfrac{1}{|\bar{x}-\bar{x}'|}}{\partial n} u\left(-\frac{|\bar{x}-\bar{x}'|}{a}\right)$$
$$+ \frac{1}{|\bar{x}-\bar{x}'|}\frac{\partial|\bar{x}-\bar{x}'|}{\partial n}\int_{-t'}^{t''} u\frac{\partial F(|\bar{x}-\bar{x}'|+at)\,dt}{\partial t}$$
$$= \frac{\partial \dfrac{1}{|\bar{x}-\bar{x}'|}}{\partial n} u\left(-\frac{|\bar{x}-\bar{x}'|}{a}\right)$$
$$-\frac{1}{a}\frac{1}{|\bar{x}-\bar{x}'|}\frac{\partial|\bar{x}-\bar{x}'|}{\partial n}\frac{\partial u}{\partial t}\bigg|_{t=-(|\bar{x}-\bar{x}'|)/a} \tag{38}$$

with the last equality obtained by partial integration.

Moreover, the time integration of the second term of the first integral in (32) gives

$$\int_{-t'}^{t''} v\frac{\partial u}{\partial n} = a\int_{-t'}^{t''}\frac{F(|\bar{x}-\bar{x}'|+at)}{|\bar{x}-\bar{x}'|}\frac{\partial u}{\partial n} = \frac{1}{r_0} f\left(-\frac{|\bar{x}-\bar{x}'|}{a}\right), \tag{39}$$

where

$$f = \frac{\partial u}{\partial n}. \tag{40}$$

At this point Kirchhoff changed the initial point of the time variable so that the time, until now called 0, became the time t. With this new convention, insertion of (36), (38) and (39) into (32) led Kirchhoff to the final expression:

$$4\pi u(\bar{x}', t) = \int_{\partial\Omega} \left(\frac{\partial \frac{1}{|\bar{x} - \bar{x}'|}}{\partial n} u\left(t - \frac{|\bar{x} - \bar{x}'|}{a}\right) - \frac{1}{a}\frac{1}{|\bar{x} - \bar{x}'|}\frac{\partial |\bar{x} - \bar{x}'|}{\partial n}\frac{\partial u\left(t - \frac{|\bar{x} - \bar{x}'|}{a}\right)}{\partial t} - \frac{1}{|\bar{x} - \bar{x}'|} f\left(t - \frac{|\bar{x} - \bar{x}'|}{a}\right)\right) d\sigma \tag{41}$$

or collecting the two first terms:

$$4\pi u(\bar{x}', t) = \int_{\partial\Omega} \left(\frac{\partial}{\partial n} \frac{u\left(t - \frac{|\bar{x} - \bar{x}'|}{a}\right)}{|\bar{x} - \bar{x}'|} - \frac{f\left(t - \frac{|\bar{x} - \bar{x}'|}{a}\right)}{|\bar{x} - \bar{x}'|}\right) d\sigma. \tag{42}$$

Physically (42) says that the amplitude of the wave at a point in Ω is found as a superposition of waves propagated from the boundary with the velocity a. Thus (42) gives a mathematical expression for Huygens' principle.

Kirchhoff's method could not easily be adapted to two dimensions because it would lead to certain divergent integrals. Therefore Volterra in his work on this problem [1894] chose a different kind of auxiliary function v to be used in connection with Green's theorem.

8. Thus far we have met no general definition of a fundamental solution and no such definition was actually given before 1908 (for elliptic equations see [Hadamard 1904/05]). In that year Jacques Hadamard (1865–1963) published an article, "Théorie des équations aux dérivées partielles linéaires hyperboliques et du problème de Cauchy," in *Acta Mathematica.* He lectured on the subject in 1920 at Yale University; these lectures were published in 1922 in English and later translated into Hadamard's native tongue [1932].

S. Mandelbrojt and L. Schwartz wrote in an obituary note [1965] on Hadamard:

> This book [Hadamard 1932] is a real masterpiece and by its content, its clarity, and the abundance of its ideas, it has inspired all the investigators on partial differential equations of the following generation.

The book contains three of Hadamard's most ingenious contributions to analysis: The definition of a well-posed boundary value or initial value problem, the definition of a fundamental—or as he called it—elementary solution, and the partie finie of a divergent integral. I shall comment on the last two ideas mentioned.

Concerning the aim of the book Hadamard wrote [Hadamard 1932, Preface]:

> Je me suis proposé de poursuivre le travail du géomètre italien [Volterra] et pour cela de le modifier et de l'étendre de sorte qu'il devienne applicable à toutes les équations hyperboliques (normales) ou lieu de l'être à une seule d'entre elles.

Hadamard found Volterra's form of the auxiliary function v inadequate because it was hard to generalize to equations other than the wave equation and because it was not independent of changes of coordinates. Nor could he use Kirchhoff's method. So in order to define appropriate "elementary solutions" he chose Green's function (20) for an elliptic equation (19) as his starting point:

> Il est évident que, si nous restons dans le cas analytique, il n'existe pas de différence essentielle entre l'équation (19) et l'équation hyperbolique de Laplace
>
> $$F(u) = \frac{\partial^2 u}{\partial x\, \partial y} + A\frac{\partial u}{\partial x} + B\frac{\partial u}{\partial y} + Cu = 0 \tag{43}$$
>
> (A, B, C étant encore des fonctions de x et y), que l'on déduit évidemment de la première, en raison de la présence d'un infini d'orde $\frac{3}{2}$ le long de la frontière, la $(x - x_0)^2 + (y - y_0)^2$ en $(x - x_0)(y - y_0)$, on doit trouver, pour (43), une solution de la forme
>
> $$U \log [(x - x_0)(y - y_0)] + W. \tag{44}$$
>
> [Hadamard 1932, p. 100.]

Hadamard could prove that the regular function U was the Riemann function corresponding to the adjoint of (43) (i.e., U satisfied (23)), thus establishing the connection between Green's and Riemann's functions.

The solution (44) was generalized by Hadamard to the higher-dimensional equations:

$$F(u) = \sum_{i,k=1}^{m} A_{ik}\frac{\partial^2 u}{\partial x_i\, \partial x_k} + \sum_{i=1}^{m} B_i\frac{\partial u}{\partial x_i} + Cu = f. \tag{45}$$

He noted that in the case (43)

$$\Gamma(x, y, x_0, y_0) = (x - x_0)(y - y_0) = 0 \tag{46}$$

was the normal equation of the characteristic conoid with vertex at (x_0, y_0). This led him to the more general definition of an elementary solution to (45):

$$\begin{aligned} v &= \frac{\mathscr{U}(\bar{x}, \bar{x}_0)}{\Gamma(\bar{x}, \bar{x}_0)^{(m-2)/2}} \quad \text{for } m \text{ uneven,} \\ v &= \frac{\mathscr{U}(\bar{x}, \bar{x}_0)}{\Gamma(\bar{x}, \bar{x}_0)^{(m-2)/2}} - U \log \Gamma(\bar{x}, \bar{x}_0) \quad \text{for } m \text{ even,} \end{aligned} \tag{47}$$

the first term of which corresponds to (13(b)). Again in (47) $\Gamma(\bar{x}, \bar{x}_0) = 0$ is the normal equation of the characteristic conoid with vertex at x_0 and the functions $\mathcal{U}$ and U are holomorphic functions having a specified value at $\bar{x}_0$. Hadamard proved the existence of the elementary solution to (45) [Hadamard 1932, p. 147] and applied it in a way similar to that of Green and Riemann to solve the corresponding Cauchy problem. The discussion of this point can be found in §11.

Formula (47) gives a general definition of the elementary solutions of second-order hyperbolic or elliptic equations (in the elliptic case, complex characteristics must be used). However, in the hyperbolic case, Hadamard's elementary solutions are not fundamental solutions in the modern sense, i.e. they do not satisfy (1). For instance, for the wave equation in physical space ($m = 4$), the fundamental solution (30) as found by Kirchhoff has support on the light cone only, whereas the support of (47) exceeds this set (cf. [Schwartz 1950/51, Vol. I, p. 134]).

9. Hadamard's definition of the elementary solution was the most influential one before the creation of the theory of distributions, but it was not the only one. An alternative definition was given in 1911 by the Swede Niels Zeilon who began his article, "Das Fundamentalintegral der Allgemeinen Partiellen Linearen Differentialgleichungen mit konstanten Koeffizienten" [1911], with the words:

> Für die folgenden Ausführungen ist es zwechmässig den Begriff des Fundamentalintegrales in etwas anderer Weise zu fixieren, als man das gewöhnlich tut. Es soll jede Funktion $F(x, y, z)$ ein Fundamentalintegral der linearen Differentialgleichung
>
> $$f\left(\frac{\partial}{\partial x}, \frac{\partial}{\partial y}, \frac{\partial}{\partial z}\right)u = 0 \tag{48}$$
>
> genannt werden, die der Bedingung genügt, dass
>
> $$f\left(\frac{\partial}{\partial x}, \frac{\partial}{\partial y}, \frac{\partial}{\partial z}\right) \int_D F(x - \lambda, y - \mu, z - \nu)\, d\lambda\, d\mu\, d\nu \tag{49}$$
>
> gleich 1 ist, wenn das Integrationsgebiet D den Punkt x, y, z einschliesst, und gleich 0, wenn dieser Punkt ausserhalb D liegt. Oder was auf dasselbe herauskommt: Wenn $\varphi(x, y, z)$ eine willkürliche Funktion ist, so soll
>
> $$u = \int_D F(x - \lambda, y - \mu, z - \nu)\varphi(\lambda, \mu, \nu)\, d\lambda\, d\mu\, d\nu \tag{50}$$
>
> im Gebiete D eine Lösung geben der Gleichung
>
> $$f\left(\frac{\partial}{\partial x}, \frac{\partial}{\partial y}, \frac{\partial}{\partial z}\right)u = \varphi. \tag{51}$$
>
> Dabei ist D als ganz willkürlich vorausgesetzt, namentlich muss es gestattet sein, es beliebig klein zu machen.

If one formally moves the differentiation in (49) inside the integral sign one sees that Zeilon is describing the δ-behavior of

$$f\left(\frac{\partial}{\partial x}, \frac{\partial}{\partial y}, \frac{\partial}{\partial z}\right)F.$$

This gives a direct link to Schwartz' definition (1). More directly the last descriptions (50) and (51) of F are identical with (5) and (4). Therefore it is clear that Zeilon's definition of the fundamental solution is equivalent to the modern one. Zeilon thus skillfully avoided the direct use of the δ-function in the definitions of the fundamental integral. Yet from what we have seen concerning the wave equation (30), (31), it is obvious that he would need the δ-function if he had to write the expression for the fundamental solution to the wave equation. However he avoided this by altering his definition in such cases:

> ... immerhin dürfte man sagen können, dass die Lösung der Differentialgleichung entweder durch den normalen Ausdruck [i.e. (50)] gegeben wird, oder, falls dieser divergieren sollte, durch
>
> $$u = g\left(\frac{\partial}{\partial x}, \frac{\partial}{\partial y}, \ldots\right)\int G(x - \lambda, y - \mu, \ldots)\varphi(\lambda, \mu, \ldots)\, d\lambda\, d\mu \ldots \tag{52}$$
>
> wo G für die Gleichung
>
> $$g\left(\frac{\partial}{\partial x}, \frac{\partial}{\partial y}, \ldots\right)f\left(\frac{\partial}{\partial x}, \frac{\partial}{\partial y}, \ldots\right)u = 0 \tag{53}$$
>
> gebildet ist.

Thus, instead of taking the fundamental integral F itself, Zeilon took a suitable integral $g^{-1}(\partial/\partial x, \partial/\partial y, \ldots)F$ as the essential quantity. This is very much like the trick used by Volterra [1894] who first integrated with respect to t (i.e. he took $g = \partial/\partial t$) and afterwards differentiated outside the final integral sign (cf. [Hadamard 1932, p. 163]).[8] Zeilon in his paper showed how one could construct fundamental solutions using Fourier integral techniques, but as far as I know, it was not very influential.

10. In the development of the fundamental solutions as it has been described here, the δ-function has played two roles: as the right-hand side of (1) and as the singularity of the fundamental solution of the wave equation (30). Except for §6 on Kirchhoff this part can be viewed as the story of the different ways the δ-function was circumvented. But from another point of view it tells how the intuitive idea of (1) was rigorously represented in the mathematics of the day, in fact often in a way which anticipated the solution of the problem in the theory of distributions. For instance, Green's proof that

$$\int \Delta\left(\frac{1}{|\bar{x} - \bar{x}'|}\right)V(x) = V(x') \tag{54}$$

(cf. (15)) anticipated the functional behaviour of $\delta(\bar{x} - \bar{x}')$; Courant–Hilbert's introduction of the approximating δ-function (9) hints at the sequence definition of distributions; and Zeilon's definition (50) + (51) is close to a definition of δ as a convolution operator δ^*:

$$\delta^*\varphi = \varphi \tag{55}$$

a definition which Schwartz later used (see Ch. 6, §6). Thus the way was paved for Schwartz' definition (1) which has now become the universally accepted definition.

11. The δ-function and the generalized functions derived from it are studied in further detail in the latter part of this chapter (§16–49). These "functions" constituted the most important class of generalized functions before the development of the theory of distributions. However with our present-day knowledge of distributions, we can distinguish another class of early generalized functions: expressions involving Hadamard's partie finie (finite part). In contrast to the δ-function Hadamard's partie finie was not in its early stage recognized as a generalized function but only as a generalized integral. The two groups of generalized objects also differed with respect to rigor. Whereas the δ-function was crying out for a solid basis, the partie finie rested from the start on a sound foundation, created as it was by one of the greatest mathematicians of the time, Jacques Hadamard. His definition of the partie finie was directly motivated by his work with the Cauchy problem for equation (45). In this case, Green's formula took the form

$$\int_{\Omega} [vF(u) - uF'(v)]\, da = -\int_{\partial\Omega} \left(v\frac{du}{d\nu} - u\frac{dv}{d\nu} + Luv \right) d\sigma, \tag{56}$$

where L is a certain function of x_i ($i = 1, \ldots, m$) and ν is the so-called transversal direction (the specific form of L and ν is of no interest in this connection). In order to solve the equation (45) with the Cauchy conditions:

$$u = u_0, \qquad \frac{\partial u}{\partial \nu} = u \tag{57}$$

on a noncharacteristic surface S, Hadamard used a procedure which was a combination of Green's and Riemann's. He applied (56), with v equal to the elementary solution, to a domain Ω bounded (like Riemann's) by the characteristic cone with vertex $\bar{x}_0$ and the surface S. Like Green he removed a small ball with center $\bar{x}_0$, leaving the domain Ω'.

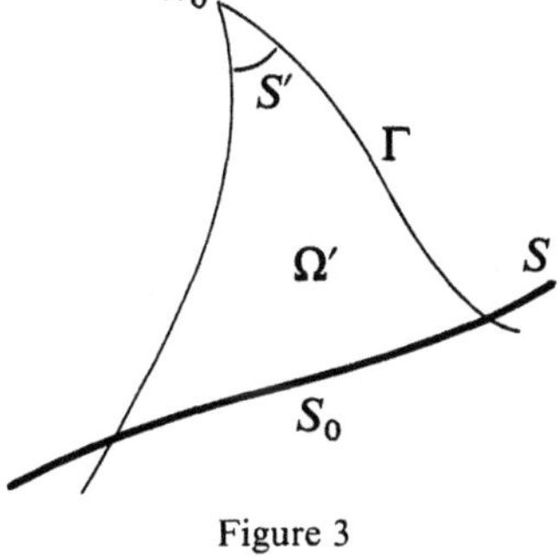

Figure 3

But contrary to what happened in the elliptic case, the elementary solution did not only have a singularity at a point but along the whole characteristic conoid, so that Hadamard was still left with two divergent integrals. He overcame this difficulty by introducing the generalized integral called the partie finie, which he denoted $\overline{\left|\int\right.}\cdots$. Nowadays it is usually written Pf $\int\cdots$. In the next section I will give his definition in some simple cases. I shall for the present take the definition for granted and formulate Hadamard's result in terms of it: from (56) he found

$$\overline{\left|\iiint_{\Omega'} vF(u)\,da\right.} = \overline{\left|\int_{\partial\Omega'} v\frac{du}{d\nu} - u\frac{dv}{d\nu} + Luv\,d\sigma\right.}. \tag{58}$$

The integral over the part of the small sphere will tend to a constant times $u(\bar{x}_0)$ when the radius tends to zero, just as it happened with Green. Therefore Hadamard obtained the final result

$$(-1)^{(m-1)/2}\pi\Omega_{m-3}f(x_0) = \overline{\left|\int_{\Omega} vf\,da\right.} + \overline{\left|\int_{\partial\Omega}\left(u_0\frac{dv}{d\nu} - vu_1 - Lu_0 v\right)d\sigma\right.}. \tag{59}$$

In the last term the partie finie means that the integral should only be taken (as a partie finie as explained below) on S_0 (the part of S belonging to $\partial\Omega$), while the surface integral over the characteristic conoid should be disregarded.

12. Before I turn to Hadamard's definition of the partie finie, let me indicate another related problem in which it proved useful—this is actually the example which Hadamard [1932, p. 165] used as the motivation for the introduction of the generalized integral:

It can be proved [Hadamard 1932, p. 71] that the solution of the two(space)-dimensional wave equation

$$\frac{\partial^2 u}{\partial x^2} + \frac{\partial^2 u}{\partial y^2} - \frac{\partial^2 u}{\partial t^2} = 0 \tag{60}$$

satisfies

$$u(x_0, y_0, t_0) = \frac{1}{2\pi}\left[\mu_{t_0}(u_1) + \frac{d}{dt_0}\mu_{t_0}(u_0)\right], \tag{61}$$

where

$$\mu_{t_0}(f) = \iint \frac{f(x, y)\,dx\,dy}{\sqrt{t_0^2 - (x - x_0^2) - (y - y_0^2)}} \tag{62}$$

integrated over the circle $(x - x_0)^2 + (y - y_0)^2 \leq t_0^2$. (Poisson had proved this result for three space dimensions).

Hadamard commented on the formula (61):

> Pour cela, la méthode usuelle consisterait à differentier par rapport à t_0, sous le signe $\iint$, ce que porterait seulement sur le dénominateur; et, d'un autre côté, à tenir

compte du fait que la limite est variable avec t_0, ce que donnerait lieu à un terme de frontière, savoir une intégrale simple le long de la circonférence. Mais il apparaît immédiatement que les intégrales double et simple sònt dépourvues de sens: la première, en raison de la présence d'un infini d'ordre $\frac{3}{2}$ le long de la frontière, la seconde parce que chacun de ses éléments est infini. Naturellement, des artifices simples permettraient de faire la différentiation en évitant cet inconvénient [In a footnote he explains how this can be done by a transformation of variables]: mais il seraient sans intérêt pour nous, car—quelque paradoxal que cela semble—notre méthode va consister à ne pas l'éviter.

13. Thus Hadamard was faced with the problem of defining the integral of a quantity tending to infinity as $1/r^{\alpha}$ at the boundary of the domain of integration. He defined it first in one dimension for $\alpha = \frac{3}{2}$:

Partial integration gives for $A \in C^1$:

$$\begin{aligned}\int_a^x \frac{A(y)}{(b-y)^{3/2}}\,dy &= 2\left[\frac{A(y)}{(b-y)^{1/2}}\right]_a^x - 2\int_a^x \frac{A'(y)}{(b-y)^{1/2}}\,dy \\ &= 2\frac{A(x)}{(b-x)^{1/2}} + \left[-2\frac{A(a)}{(b-a)^{1/2}} - 2\int_a^x \frac{A'(y)}{(b-y)^{1/2}}\,dy\right].\end{aligned} \tag{63}$$

For x tending to b, the first term will tend to infinity (the infinite part) and the terms in the square bracket will tend to a fixed finite value which Hadamard called the finite part of the integral.[9] He denoted it

$$\overline{\left|\int_a^b \frac{A(y)}{(b-y)^{3/2}}\,dy.\right.} \tag{64}$$

In the same way he showed that for other powers $(p+\mu)(p \in \mathbb{N} \cup \{0\}$ and $0 < \mu < 1)$ of $(b-y)$ and for $A(y) \in C^p$ there existed a function $B(x) \in C^p$ such that

$$\int_a^x \frac{A(y)}{(b-y)^{p+\mu}}\,dy + \frac{B(x)}{(b-x)^{p+\mu-1}} \tag{65}$$

had a limiting value for x tending to b. This value which is unique he correspondingly called the partie finie of $\int_a^x [A(y)/(b-y)^{p+\mu}]\,dy$.

For expressions of the form

$$\int_a^b \frac{A(x)}{(b-x)^p}\,dx \tag{66}$$

for $p \in \mathbb{N}$, the partie finie was similarly obtained by adding an expression of the form

$$\frac{B(x)}{(b-x)^{p-1}} + B_1(x)\log(b-x). \tag{67}$$

However, for $p \geq 2$ addition of a term $K(b-x)^{p-1}$ to B could alter the resulting limit by a constant K. For such ps therefore the partie finie was not determined. Hadamard remarked that this could be overcome by specifying of the added term, but he made no such convention to avoid the ambiguity. (In *Théorie des Distributions* Schwartz specified that $B(x)$ be a polynomial of degree $p - 2$.) Following the definition Hadamard proved several rules for calculation with the partie finie of divergent integrals. For example, he showed how to differentiate with respect to the upper limit b of the integral:

$$\frac{d}{db}\overline{\left|\int_a^b \frac{A(x)}{(b-x)^{1/2}}\,dx\right.} = \overline{\left|\int_a^b -\frac{1}{2}\frac{A(x)}{(b-x)^{3/2}}\,dx.\right.} \tag{68}$$

This result, which is the one-dimensional analogue of the problem mentioned in §12 can be obtained formally by differentiating under the integral sign and omitting a term corresponding to the upper boundary.

Hadamard generalized the notion of partie finie to multiple integrals of the form

$$\iiint_T \frac{A(x, y, z)}{G(x, y, z)^{p+\alpha}}, \tag{69}$$

where part of the boundary ∂T of the domain of integration is composed of the surface $G(x, y, z) = 0$. As we have already seen, this multi-dimensional partie finie gave Hadamard the solution (59) to the Cauchy problem. After having given the analysis leading to the formula (59), he proved synthetically that it satisfied the equation and the boundary values. Malgrange wrote [Lévy *et al.* 1967, p. 45]:

> Il convient de noter ici, combien cette notion pouvait sembler inattendue et paradoxale à l'époque, et à Hadamard lui-même, comme il l'a dit souvent.

14. Hadamard's introduction of the partie finie was done in a very *ad hoc* way.[10] However, it proved to be of a more fundamental nature when seen in the light of the theory of distributions.

In his first article on the theory of distributions Schwartz [1945, pp. 62–63] showed how the partie finie naturally arose as a so-called pseudo-function:

> La dérivée de la fonction $f(x)$, null pour $x \leq 0$, égale à $1/\sqrt{x}$ pour $x > 0$, n'est certainement pas la fonction dérivée usuelle, nulle pour $x \leq 0$, àgale $\bar{y} - \frac{1}{2}x^{-3/2}$ pour $x > 0$, car cette fonction n'étant pas sommable au voisinage de $x = 0$, ne définit pas une distribution.
>
> Le calcul direct donne:
>
> $$\begin{aligned} f'(\varphi) = -f(\varphi') &= -\int_0^{+\infty} \frac{\varphi'(x)}{\sqrt{x}}\,dx = -\lim_{\varepsilon\to 0}\int_\varepsilon^{+\infty} \frac{\varphi'(x)}{\sqrt{x}}\,dx \\ &= \lim_{\varepsilon\to 0}\left[-\int_\varepsilon^{+\infty} \frac{1}{2}x^{-3/2}\varphi(x)\,dx + \frac{\varphi(0)}{\sqrt{\varepsilon}}\right]. \end{aligned} \tag{70}$$

On obtient donc ce que *M*. Hadamard a appelé la "partie finie" d'une intégrale divergente et qu'il a noté par le symbole:

$$f'(\varphi) = \left\lceil \int_0^{+\infty} \left(-\frac{1}{2}x^{-3/2}\right)\varphi(x)\,dx. \right. \tag{71}$$

De sorte que nous pouvons considérer que f' est une pseudo-fonction que nous représentons par 0 pour $x \leq 0$, et par

$$\left\lceil -\frac{1}{2}x^{-3/2} \right. \quad \text{pour } x > 0. \tag{72}$$

On démontrerait de même, en utilisant le symbole $\lceil\;$ de M. Hadamard, que la dérivée d'ordre k de la fonction nulle pour $x \leq 0$, égale à x^α pour $x > 0$, $(\alpha > -1)$ est nulle pour $x \leq 0$, et représentée par la pseudofonction:

$$\lceil \alpha(\alpha - 1)\ldots(\alpha - k + 1)x^{\alpha-k} \quad \text{pour } x > 0. \tag{73}$$

La dérivée de la fonction log $|x|$ est la psuedo-fonction

$$\lceil 1/x = \text{v.p. } 1/x, \tag{74}$$

v.p. désignant la valeur principale de Cauchy

$$\text{v.p. } \frac{1}{x}(\varphi) = \text{v.p.} \int_{-\infty}^{+\infty} \frac{\varphi(x)}{x}\,dx = \lim_{\varepsilon \to 0} \left(\int_{-\infty}^{-\varepsilon} + \int_{\varepsilon}^{+\infty}\right). \tag{75}$$

Les dérivées des fonctions usuelles donnent des types d'intégrales généralisées d'une richesse très grande comme peuvent le montrer ces quelques exemples.

The relationship between Cauchy's valeur principale and the partie finie was already known by Hadamard. However, Malgrange pointed out [Levi *et al.* 1967] that for the valeur principale "il n'était pas nécessaire de retrancher des infiniments grands, et, pour cette raison, la 'partie finie' semblait à Hadamard de nature fort differente".

15. How did the theory of distributions influence Hadamard's partie finie? The term, which was originally attached to integrals became linked with functions. Thereby it lost its mysterious and *ad hoc* character and became a naturally integrated part of a more thorough generalization.

Conversely, how did the partie finie contribute to the invention of the theory of distributions? Apparently it did not contribute directly. Laurent Schwartz knew about the partie finie but, according to his own testimony [Schwartz 1978, Interview] it did not serve as a source of inspiration for him when he created the distributions. Nor did it inspire other mathematicians to similar theories. Neither Sobolev, Bochner nor Carleman saw the connection between their distribution-like theories and Hadamard's notion. However, it was certainly of support to the theory of distributions, once invented, that the generalized derivatives of $(1/x)^\alpha\chi(\mathbb{R}_+)$ were described so easily in terms of Hadamard's familiar term.

Part 2. The δ-function

16. The physical significance of the δ-function, as a representation of point masses, is so great that the idea underlying this earliest generalized function can be traced nearly as far back as one wishes. Already in ancient Greece and in the Middle Ages the atomic versus the continuous structure of matter was discussed. However it would be misleading to interpret these disputes as a question of the existence of a function—the mass distribution—supported by one point but having a finite integral representing the mass of the particle. In the following, therefore, I shall restrain from such excessively implicit ideas about the δ-distribution and concentrate on the cases which involve more mathematical considerations.

It is remarkable how successfully physicists and mathematicians (as we saw in the last part) have been able to avoid the mathematically troublesome δ-function in various places where it seems to us to play a fundamental role. For instance, in the treatment of gravitational or electrical forces, the traditional procedure has been to treat point masses (charges) first and then to make the transition to continuous mass or charge distributions in space. In this theory, however, the original point masses or charges together with line and surface distributions could not be treated. This apparent logical contradiction, which has been, and usually still is, involved in the argument, is the price physicists had to pay when they did not use the δ-function.

The δ-function must have had a very sad childhood since neither mathematicians nor physicists recognized it as belonging to their domain. If mathematicians used it, it was as an intuitive physical notion with no mathematical reality, as for example Courant–Hilbert used it (§2).[11] On the other hand, physicists usually considered the δ-function, or the point mass as a pure mathematical idealization which did not exist in nature.[12] Maxwell in *A Treatise on Electricity and Magnetism* [1873] commented on the formula for the potential of a point charge:

$$V = \frac{e}{r}, \qquad r = ((x-a)^2 + (y-b)^2 + (x-c)^2)^{1/2} \tag{76}$$

in the following way [§129, see also §81]:

> Hence, the value of V, as given by equation (76), may be the actual value of the potential in the space outside a closed surface surrounding the point (a, b, c), but we cannot, *except for purely mathematical purposes*, suppose this form of the function to hold up to and at the point (a, b, c) itself. For the resultant force close to the point would be infinite, ..., it would require an infinite expenditure of work to charge a point with a finite quantity of electricity.[13] (My italics.)

Thus for Maxwell the physical reality of point charges was "small bodies of which the dimensions are negligible compared with the principal distances concerned".

The situation preceding the discovery of the theory of distributions therefore seems paradoxical: by introducing the δ-function into the mathematical treatment of physical problems, the physical system which had a perfectly rigorous mathematical model was approximated by a nonexisting idealization, for which no rigorous mathematical model existed (this does not apply to quantum mechanics). Why then did physicists make such approximations; why did they not adhere to the situation existing in nature and its mathematical model? One reason is that the exact physical situation is usually unknown so that some approximation must be made. For instance in the case of a point mass or point charge, the shape of the atom (the ion or electron) was unknown. But why choose this special approximation? Because it made calculations much easier than any other approximation. I have illustrated this in note 14 by discussing two instances from the beginning of this century in which the δ-function entered as an approximation, without being explicitly defined.

17. Not only the δ-function but also its derivatives appeared naturally in mathematical physics. "Après avoir beaucoup réfléchi à cette question", Poisson in [1821/22, pp. 254 and 263] defined an "élément magnétique" as an "extrèmement petit" part of matter in which the two magnetic fluids are "très peu écartés les uns des autres" and in which the amount of each fluid is the same. He showed how the experimental results in magnetostatics could be explained on the assumption that magnetism was produced by such elements. Maxwell discussed dipoles in a more mathematical way. In [1873, §129] he explained:

> A point of first degree may be supposed to consist of two points of degree zero [point charges], having equal and opposite charges M_0 and $-M_0$, and placed at the extremities of the axis h. The length of the axis is then supposed to diminish and the magnitude of the charges to increase, so that their product $M_0 h$ is always equal to M_1. The ultimate result of this process when the two points coincide is a point of the first degree, whose moment is M_1 and whose axis is h_1.

Letting such points of first degree or dipoles approach each other along another axis h_2, Maxwell obtained quadruple points and so on. He proved that the resulting potentials were of the form

$$(-1)^n M_i \frac{\partial^i}{\partial h_1\, \partial h_2 \cdots \partial h_i} \frac{1}{r}. \tag{77}$$

The corresponding distribution of charge of course corresponds to

$$M_i \frac{\partial^i}{\partial h_1\, \partial h_2 \cdots \partial h_n} \delta.$$

This last part of Chapter 4 falls into two subparts. The first subpart (§18–33) shows how the δ-function and distributions derived from it were *used* in different mathematical and physical contexts, such as the proof of Fourier's integral theorem, electrical engineering, Fourier transformation theory, and quantum mechanics. The second subpart (§34–49) traces the attempts to give the δ-function a rigorous foundation; among these will be a "theory of distributions" developed by the Dutch physicist Tolhoek.

I do not pretend to give an account of all applications of, or foundational questions about, the δ-function. I only attempt to give a picture of the unstructured but lively activity involving this first generalized function.

18. A mathematical expression and description of the δ-function most likely appeared for the first time in connection with Fourier series and integrals, as early as 1822 in Fourier's book *Théorie Analytique de la Chaleur.*[15] It entered the discussion in two places: first, Fourier proved how the Fourier series expansion could lead to an integral expression

$$f(x) = \int \delta(\alpha - x) f(\alpha), \tag{78}$$

with

$$\delta(x) = \frac{1}{\pi}\left(\frac{1}{2} + \sum_{i=1}^{\infty} \cos ix\right) \quad \text{for } x \in [-\pi, \pi]; \tag{79}$$

second, he showed conversely how expressions like (79) could be used to "prove" the "convergence" of the Fourier integrals. The first mention can be found in Fourier [1822, §235, 3°]. In the Fourier expansion

$$\pi F(x) = \frac{1}{2}\int F(\alpha)\, d\alpha + \begin{cases} \cos x \int F(\alpha) \cos \alpha \, d\alpha + \cos 2x \int F(\alpha) \cos 2\alpha \, d\alpha \\ \qquad + \cdots + \\ \sin x \int F(\alpha) \sin \alpha \, d\alpha + \sin 2x \int F(\alpha) \sin 2\alpha \, d\alpha \\ \qquad + \cdots . \end{cases} \tag{80}$$

Fourier, without making a scruple of it, interchanged the order of integration and summation and found

$$F(x) = \frac{1}{\pi}\int_{-\pi}^{\pi} F(\alpha)\, d\alpha \begin{cases} \frac{1}{2} + \cos x \cos \alpha + \cos 2x \cos 2\alpha + \cdots \\ \quad + \sin x \sin \alpha + \sin 2x \sin 2\alpha + \cdots \end{cases} \tag{81}$$

or

$$F(x) = \frac{1}{\pi}\int_{-\pi}^{\pi} F(\alpha)\left(\frac{1}{2} + \sum_{i=1}^{\infty} \cos i(x - \alpha)\right) d\alpha. \tag{82}$$

> L'expression $\frac{1}{2} + \sum \cos i(x - \alpha)$ représente une fonction de x et de α telle que, si on la multiplie par un fonction quelconque $F(\alpha)$ et si, après avoir écrit $d\alpha$, on intègre entre les limites $\alpha = -\pi$ et $\alpha = \pi$, on aura changé la fonction proposée $F(\alpha)$ en une pareille fonction de x, multipliée par la demi-circonférence π.

In a footnote in *Oeuvres de Fourier* Darboux remarked:

Plus exactement, $F(x)$ est la limite de l'expression

$$\frac{1}{\pi}\int F(\alpha)\,d\alpha\left[\sum_{i}^{p}\cos i(x-\alpha)+\tfrac{1}{2}\right]^{16} \tag{83}$$

lorsque p augmente indéfiniment. La série

$$\tfrac{1}{2}+\cos(x-\alpha)+\cos 2(x-\alpha)+\cdots \tag{84}$$

ayant une somme indéterminée, on ne peut attacher aucun sens à l'expression

$$\tfrac{1}{2}+\sum_{i=1}^{\infty}\cos i(x-\alpha) \tag{85}$$

considérée par Fourier.

This is the typical judgment of a rigorist from the last part of the nineteenth century. From the point of view of classical analysis he is right, but in the theory of distributions Fourier's argument makes perfect sense since

$$\frac{1}{2}+\sum_{i=1}^{p}\cos i(x-\alpha)\to\pi\sum_{n=-\infty}^{\infty}\delta(x+n\cdot 2\pi), \tag{86}$$

when the convergence is taken in $\mathscr{D}'$ [Schwartz 1950/51, Vol. II, p. 82].

19. An argument resembling the converse can be found in Fourier [1822, §415 and §423] where Fourier attempted a "convergence" proof for the Fourier integral theorem

$$f(x)=\frac{1}{\pi}\int_{-\infty}^{\infty}f(\alpha)\,d\alpha\int_{0}^{\infty}\cos p(\alpha-x)\,dp^{17} \tag{87}$$

and theorem (80), respectively. In the proof of (87) he first performed the p integration over an interval $[0, p]$ and found

$$f(x)=\frac{1}{\pi}\int_{-\infty}^{\infty}f(\alpha)\,\frac{\sin p(\alpha-x)}{\alpha-x}\,d\alpha. \tag{88}$$

On doit donc donner à p, dans cette dernière expression, une valeur infinie.

In order to prove (88) Fourier used the well-known identity

$$\int_{0}^{\infty}\frac{\sin x}{x}\,dx=\frac{\pi}{2} \tag{89}$$

to conclude that (Fourier had exchanged x and α):

$$\int_{0}^{\infty}\frac{\sin px}{x}\,dx=\frac{\pi}{2} \tag{90}$$

for all positive values of p, finite as well as infinite. He then argued that for $p = \infty$ the only positive contribution to (90) comes from an infinitely small interval around zero:

> Les sinuosités de la courbe dont sin px/x est l'ordonnée sont infiniment voisines. Leur base est une longueur infiniment petite, égale à π/p. Cela étant, si l'on compare l'aire positive qui repose sur un de ces intervalles π/p à l'aire négative qui repose sur l'intervalle suivant, et si l'on désigne par X l'abscisse finie et assez grande qui répond au commencement du premier arc, on voit que l'abscisse x, qui entre comme dénominateur dans l'expression sin px/x de l'ordonnée, n'a aucune variation sensible dans le double intervalle $2\pi/p$ qui sert de base aux deux aires. Par conséquent, l'intégrale est la même que si x était une quantité constante. Il s'ensuit que la somme des deux aires qui se succèdent est nulle.
>
> Il n'en est pas de même lorsque la valeur de x est infiniment petite, parce que l'intervalle $(2\pi/p)$ a, dans ce cas, un rapport fini avec la valeur de x. On connait par là que l'intégrale
>
> $$\int_0^{+\infty} \frac{\sin px}{x}\,dx, \tag{91}$$
>
> dans laquelle on suppose p un nombre infini, est entièrement formée de la somme de ses premiers termes, qui répondent à des valeurs extrèmement petites de x. Lorsque l'abscisse a une valeur finie X, l'aire ne varie plus, parse que les parties qui la composent se détruisent deux à deux alternativement. Nous exprimerons ce resultat en écrivant
>
> $$\int_0^{\infty} \frac{\sin px}{x}\,dx = \int_0^{\omega} \frac{\sin px}{x}\,dx = \frac{\pi}{2}. \tag{92}$$
>
> La quantité ω, qui désigne la limite supérieure de la seconde intégrale, a une valeur infiniment petite; et la valeur de l'intégrale est la même lorsque cette limite est ω et lorsqu'ellel est ∞.

With a translation of the variable Fourier obtained:

$$\int_{-\infty}^{\infty} \frac{\sin p(\alpha - x)}{\alpha - x}\,dx = \int_{\alpha-\omega}^{\alpha+\omega} \frac{\sin p(\alpha - x)}{\alpha - x}\,dx = \pi. \tag{93}$$

Finally, implicitly using the continuity of f he arrived at (88):

$$\begin{aligned}\frac{1}{\pi}\int_{-\infty}^{\infty} f(x)\,\frac{\sin p(\alpha - x)}{\alpha - x} &= \frac{1}{\pi}\int_{\alpha-\omega}^{\alpha+\omega} f(x)\,\frac{\sin p(\alpha - x)}{\alpha - x}\,dx \\ &= \frac{f(\alpha)}{\pi}\int_{\alpha-\omega}^{\alpha+\omega} \frac{\sin p(\alpha - x)}{\alpha - x} = f(\alpha).\end{aligned} \tag{94}$$

Thus translated into modern notation Fourier argued that

$$\lim_{p\to\infty} \frac{1}{\pi}\,\frac{\sin p(\alpha - x)}{\alpha - x} = \delta(\alpha - x), \tag{95}$$

a formula which is correct in $\mathscr{D}'$.[18]

Equivalently in the case of the Fourier series (80) he transformed the kernel (85) into

$$\lim_{j\to\infty}\left[\sum_{i=-j}^{j}\cos jr\right]=\lim_{j\to\infty}\left[\cos jr+\sin jr\,\frac{\sin r}{1-\cos r}\right]^{16}\qquad r=\alpha-x \quad (96)$$

and argued as before that the cosine contribution vanished in the limit, whereas

$$\frac{1}{\pi}\sin jr\,\frac{\sin r}{1-\cos r}\to\delta(r)\qquad r\in[-\pi,\pi]. \quad (97)$$

One may conclude that Fourier's formulas make perfect sense in the language of distribution theory. However, the arguments obviously do not prove the basic limit theorems (86), (95) and (97). In order to prove these formulas in $\mathscr{D}'$, one has to devise a classical convergence proof along the line indicated by Darboux in the above-mentioned footnote. This was done by Dirichlet [1829 and 1837].

20. The idea of "proving" Fourier's integral theorem or the convergence of the Fourier series by showing that the Dirichlet kernel or a similar kernel (95) tends to the δ-function can be found in many early works on Fourier series or Fourier integrals: for instance, Cauchy [1823b, p. 279 and 1827, note VI] and Poisson [1815]. The two memoirs mentioned last here were read before the Academy in Paris in 1815, i.e. prior to the publication of Fourier's proof. They both used arguments which in distribution language would be

$$\frac{1}{2}\frac{\alpha}{\alpha^2+(x-a)^2}\to\delta(x-a),\quad \text{for } \alpha\to 0. \quad (98)$$

Of course, one can still recognize approximations of the δ-function in the so-called singular integrals[19] which after Dirichlet were used to prove Fourier's theorem, but as the limit was taken outside the integral sign the use of improper functions was only implicit. Even after Dirichlet had given the rigorous proof, the δ-function continued to appear in this connection.

21. Oliver Heaviside was probably the one who most explicitly pointed out the connection between Fourier series and δ-functions. He started to work with the δ-function in the article "On operators in physical mathematics" [1893, p. 510] and extended its use in *Electromagnetic Theory*, Vols. II and III [1899 and 1912] in which he introduced the δ-function while treating a wire without self-induction [Heaviside 1899, §249].

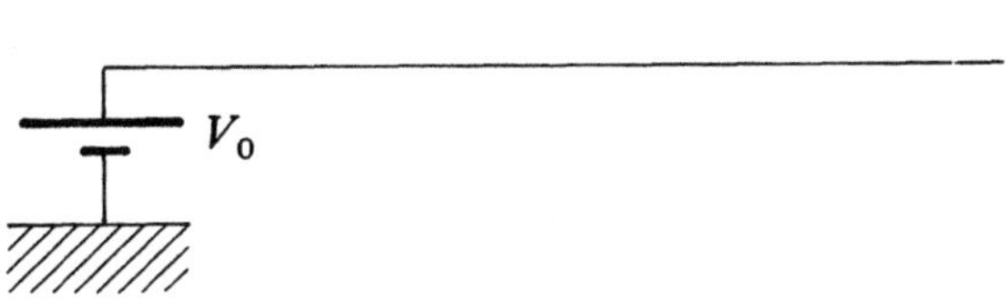

Figure 4

At time $t = 0$ an electro motive force (e.m.f.) of the form

$$V_0(t) = Q\left(\frac{R}{S\pi t}\right)^{1/2} \tag{99}$$

is impressed at the end of the cable. Here S is the permittance and R the resistance per unit length of the cable and Q is a constant charge. Heaviside had deduced, in his usual experimental way [1899, Ch. 7] that

$$p^{1/2}H(t) = (\pi t)^{-1/2},$$

where $H(t)$ is the Heaviside function

$$H(t) = \begin{cases} 1 & \text{for } t \geq 0, \\ 0 & \text{for } t < 0, \end{cases}$$

and p is Heaviside's notation for $\partial/\partial t$. Keeping this in mind (99) can be given the alternative form:

$$V_0 = (RSp)^{1/2}Q/SH(t).$$

Heaviside denoted the Heaviside function 1 and often omitted it writing

$$V_0 = (RSp)^{1/2}Q/S. \tag{100}$$

Being of the form: an expression in p operating on the suppressed Heaviside function (100) is what Heaviside called an "operational" expression. He devised three different methods of "algebrizing" such operational expressions, that is transforming them into ordinary functions of t, such as (99). For a more comprehensive treatment of Heaviside's operational calculus the reader is referred to [Lützen 1979]. In that paper [pp. 166–167] I have also explained how Heaviside in §238–242 of [1899] found that the current $C_0(t)$ at the end of the cable is operationally given by

$$C_0\left(\frac{Sp}{R}\right)^{1/2} V_0.$$

With the expression (99) or (100) for V_0 this gives:

$$C_0 = pQ = pQH(t). \tag{101}$$

"Since Q is constant for any finite value of time the result is zero", was Heaviside's first reaction to (101). Yet he continued in his polemic style:

> Is this nonsense? Is it an absurd result indicating the untruthworthy nature of the operational mathematics, or at least indicative of some modification of treatment being desirable? Not at all.... We have to note that if Q is any function of the time, then pQ is its rate of increase. If, then, as in the present case, Q is zero before and constant after $t = 0$, pQ is zero except when $t = 0$. It is then infinite. But its total amount is Q. That is to say, $p1$ means a function of t which is wholly concentrated at the moment $t = 0$, of total amount 1. It is impulsive, so to speak. The idea of an impulse is well known in mechanics, and it is essentially the same here. Unlike the function $p^{1/2}1$ [$1 = H(t)$] the function $p1$ does not involve appeal either to experiment or to generalized differentiation, but involves only the ordinary ideas of differentiation and integration pushed to their limit.

Heaviside's preceding introduction of the δ-function is much more explicit than Fourier's was. Fourier showed how certain expressions (79), (95) and (97) had the same "spotting property". However, it is not clear whether he considered them as one and the same object (at least restricted to $[-\pi, \pi]$). Heaviside, on the other hand, considered the impulsive function as a well-determined object which could be represented in different ways. Heaviside was thus the second person—Kirchhoff was the first—to give a mathematical definition of the δ-function, but a nonrigorous one.

22. Heaviside obtained the equivalent of (79) in his typically experimental way [Heaviside 1899, §267]. He found the voltage at a point x and at a time t due to a charge Q initially imposed at a point y of a cable of length l grounded at both ends to be given operationally by

$$V = \frac{\sinh qy}{\sinh ql} ql \sinh q(l - x) \frac{Q}{Sl}, \tag{102}$$

where $q^2 = RSp$. "Algebrizing" (102) he obtained the ordinary solution:

$$V = \frac{2Q}{Sl} \sum_{n=1}^{\infty} e^{p_n t} \sin s_n y \sin s_n x, \tag{103}$$

where

$$s_n = n\pi/l \quad \text{and} \quad s_n^2 = -RSp_n.$$

For $t = 0$ he knew that V represented the potential due to the point charge Q at y so for $Q/S = 1$, V would be an impulsive function, which Heaviside usually called u or $p1$. Thus

$$[\delta(x - y)] = u = \frac{2}{l} \sum_{n=1}^{\infty} \sin \frac{n\pi x}{l} \sin \frac{n\pi y}{l}. \tag{104}$$

He continued:

If we multiply a continuous function of y, say $f(y)$, by u, an impulsive function which exists only at the point $y = x$ the product is obviously zero except at that point, where it is infinite. But if we take the space total of the product $uf(y)$, the result is $f(x)$. For u only exists at x and its total is 1. Thus

$$\int uf(y)\,dy = f(x) \tag{105}$$

if the limits include the point x.

This is (78). Heaviside inserted the form (104) of u and obtained the Fourier sin series:

$$f(x) = \frac{2}{l} \sum_{1}^{\infty} \sin \frac{n\pi x}{l} \int_0^{\infty} f(y) \sin \frac{n\pi y}{l}\,dy. \tag{106}$$

Similarly, considering a cable isolated at the ends, Heaviside found [Heaviside 1899, §269]

$$[\delta(x-y)] = u = \frac{1}{l} + \frac{2}{l}\sum_{1}^{\infty} \cos\frac{n\pi x}{l}\cos\frac{n\pi y}{l} \tag{107}$$

and combining (104) and (107):

$$[\delta(x-y)] = u = \frac{1}{2l}\left\{1 + \sum_{1}^{\infty} \cos\frac{n\pi}{l}(x-y)\right\}, \tag{108}$$

the same expression Fourier found (86).

Substituting (107) and (108) into (105) gave him the cosine and the full Fourier expansion of f, and in the limit $l = \infty$ (108) gave him

$$\delta(x-y) = u = \frac{1}{\pi}\int_0^{\infty} \cos s(y-x)\, ds \tag{109}$$

(compare with Hahn's expression for the generalized Fourier transform of 1 (Ch. 3, §5)). He noticed that (104), (107) and (108) not only gave one impulse at $x = y$ but, in accordance with the periodicity of the trigonometric functions, produced a whole series of impulses, different for each expression. He illustrated this by the diagrams:

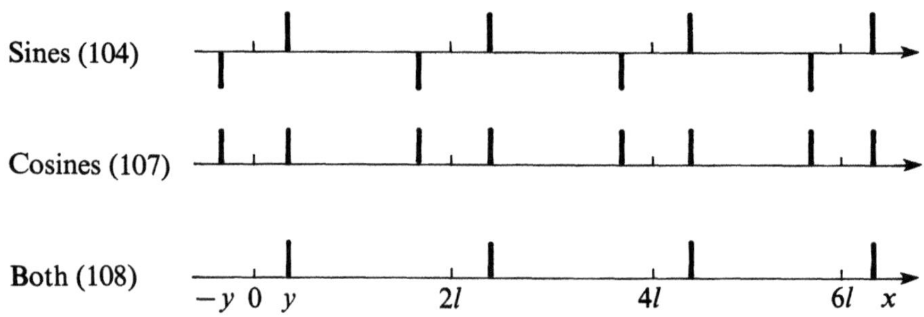

Figure 5

(compare (86)).

In this way Heaviside strengthened the connection between the Fourier series and the δ-function more than had been done during the beginning of the nineteenth century and certainly much more than the rigorous non-experimental mathematics of his time could support.[20]

23. Heaviside not only developed functions in trigonometric series, but he also gave an operational deduction of Bessel's theorem [Heaviside 1899, §526 (1)–(8)]:

$$f(t) = \iint f(x) J_m(2\sqrt{xy}) J_m(2\sqrt{yt})\, dx\, dy, \tag{110}$$

which he derived from (105) and the identity

$$u = [\delta(x - t)] = \int J_m(2\sqrt{xy})J_m(2\sqrt{yt})\,dy. \tag{111}$$

The deduction of (110) shows the operational calculus and the δ-function in their full splendor and strangeness.

Heaviside employed the notation $x_1 = d/dx$, $t_1 = d/dt$. Using series expansion and the rule $t_1^{-n} = t^n/n!$ [Lützen 1979, p. 165] he obtained

$$e^{-x_1 t}x_1 = \frac{x_1}{1 + x_1/t_1} = \frac{t_1 x_1}{x_1 + t_1} = \frac{t_1}{1 + t_1/x_1} = e^{-t_1 x}t_1. \tag{112}$$

From the operational form of Taylor's theorem, known already to Lagrange and Laplace (cf. [Koppelman 1971/72]) Heaviside got

$$e^{-t_1 x}f(t) = f(t - x). \tag{113}$$

Hence the two extreme terms of (112) are equal to u or

$$(112) = \delta(t - x) = \delta(x - t). \tag{114}$$

Therefore from (105) he found:

$$f(t) = \int_0^\infty f(x)(e^{-t_1 x}t_1)\,dx. \tag{115}$$

If we transform $f(t)$ to $f_1(t)$ then (115) becomes

$$f_1(t_1) = \int_0^\infty f(x)t_1 e^{-t_1 x}\,dx. \tag{116}$$

This is a real quantitative definite integral and its solution. That is, t_1 may be a positive variable.

In this peculiar way Heaviside derived Carson's integral equation (cf. [Lützen 1979 (III.7)]) relating the operational solution $f_1(t_1)$ of a problem to its "algebrized" solution $f(x)$. For Heaviside, however, (116) was not a means for "algebrizing" $f_1(t_1)$, a process he could do with the aid of his three algebrizing rules, but it gave a simple way of calculating the definite integrals on the right-hand side. For instance, since $t_1^{-n} = t^n/n!$, he immediately had "Euler's most valuable fundamental integral"

$$\int \frac{x^n}{n!}\,t_1 e^{-t_1 x}\,dx = t_1^{-n}. \tag{117}$$

Taking $n = 0$ this gave

$$\int e^{-(x/t_1)}\,dx = t_1. \tag{118}$$

Now let us see how he found (111) for $m = 0$. A simple series expansion of J_0 gave:

$$J_0(2\sqrt{xt}) = 1 - \frac{xt}{1!1!} + \frac{x^2t^2}{2!2!} - \cdots = 1 - \frac{x}{1!t_1} + \frac{x^2}{2!t_1^2} + \cdots$$
$$= e^{-(x/t_1)} \tag{119}$$

Therefore

$$\int_0^\infty J_0(2\sqrt{yt})J_0(2\sqrt{yx})\,dx = \int_0^\infty e^{-(y/t_1)}e^{-(y/x_1)}\,dy$$
$$= \int_0^\infty e^{-[(t_1+x_1)/(t_1x_1)]y}\,dy$$
$$= \frac{t_1x_1}{t_1+x_1} = \delta(x-t), \tag{120}$$

the last two equalities being found from (118) and (114). By a similar use of (116) Heaviside found the integral (111) for $m > 0$ and many other definite integrals, 66 in all, with many of them involving δ-functions. He said he "might go on for ever". Heaviside knew, of course, that the "Cambridge mathematicians" would not receive this method with open arms and anticipated their criticisms:

> The above [66 examples] may help others on the way. But perhaps, like the fishes who were preached to by the saint, "Much edified were they, but preferred the old way." Very well, then there let them stay.

24. In Heaviside's experimental mathematics there was also room for the derivatives of the δ-function. He started [Heaviside 1899, §253] with an impulse pQ at $t = 0$.

> Now, if this impulse be followed at time Δt later, by the impulse $-pQ$, the negative of the former, it will nearly undo the effect of the first one. If Q be finite, the differential effect will decrease with Δt, and become zero when it does. But if, whilst decreasing Δt, we simultaneously increase Q in the same ratio, the differential effect is maintained of finite size, and in the limit, where Δt vanishes and Q is infinite, whilst their product $Q\Delta t$ is finite, the result takes a special simplified finite form.

He showed that this effect could be written as p^2Q_1 where $Q_1 = Q \cdot \Delta t$ and referred to the similar introduction of magnetic moment (cf. remarks on Poisson and Maxwell, §17)

> It is clear that we may extend the above to impulses of any degree, say
>
> $$C_0 = pQ_0 \quad \text{or} \quad p^2Q_1 \quad \text{or} \quad p^3Q_2, \text{etc.}$$

Heaviside tried to use these "multiple impulses", as he called them, for the "algebrization" of

$$e = pq1 \quad \text{or} \quad p\left(\frac{RS}{\pi t}\right)^{1/2}. \tag{121}$$

Simple differentiation gave

$$p\left(\frac{RS}{\pi t}\right)^{1/2} = -\frac{1}{2t}\left(\frac{RS}{\pi t}\right)^{1/2}. \tag{122}$$

> But we must understand $pq1$ more literally. Thus, $q1$ is the function of the time which is zero before $t = 0$ and is $(RS/\pi t)^{1/2}$ later. So $pq1$ which is its rate of increase, is zero before $t = 0$ then jumps to ∞, then jumps through zero to $-\infty$, and lastly rises to zero again gradually, [according to (122)].

He did not go into more detail with this example, but his words above suggest that he thought that the derivative of $t^{-1/2}H(t)$ could be expressed as some impulse (δ or δ') and the ordinary derivative (122). On this point however his intuition failed him, for as we know now, a different type of generalized functions has to be used to solve this problem, namely the parties finies:

$$\frac{d}{dt}t^{-1/2}H(t) = \mathrm{Pf}(-\tfrac{1}{2}t^{-3/2}) \quad \text{in } \mathscr{D}'. \tag{123}$$

(cf. (19)).

25. With this examination of Heaviside's work, we have entered another scientific domain: electrical engineering, which together with quantum mechanics was the most important field of application of the δ-function prior to the theory of distributions. The widespread use of Heaviside's operational calculus from the 1920s and onwards inevitably gave rise to formal manipulating with δ-functions, for as soon as functions are to be constantly zero on the negative half-axis, the application of a positive power of $p = d/dt$ will in general introduce δ-functions. In many of the applications, however, the δ-function entered the calculations only as $p1$ or $pH(t)$ and disappeared in the result. For this reason the symbol itself was not given any special meaning (see, for instance, [Berg 1929], [Bush 1929] and [Lützen 1979, Ch. V, §1]). Sometimes however the δ-function was discussed and as late as [1946] in Josephs' *Heaviside's Electric Circuit Theory* it was described, naively, as the limit of the functions

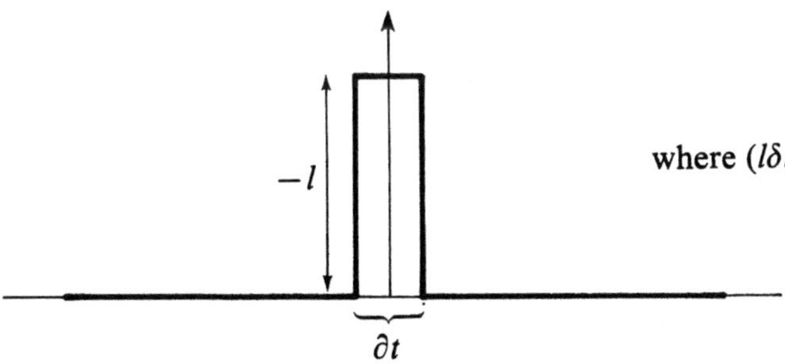

where $(l\delta t) \to 1$

Figure 6

As I have explained in [Lützen 1979] the attempts to rigorize Heaviside's operational calculus concentrated mainly on the explanation of his strange algebraical manipulation with operators and the algebrizing rules; the problems of the δ-function were neglected to a large degree. Exceptions are with Smith and Sumpner, to whom I shall return in §38–44. Therefore the different rigorized theories could only explain the success of Heaviside's procedures in cases where the δ-function did not intervene. Some of the rigorizers of the operational calculus banished the δ-function while others were not bothered by its weak foundation and used it without much comment.

26. We find both views among the persons promoting the use of the Laplace transformation in the operational calculus. Even the two persons most responsible for making this method the most successful operational procedure had different opinions concerning the δ-function; the rigorist Doetsch found it "illegitim" [1950, p. 57, footnote] and ignored it completely in his early works [1937]; the physicist and electrical engineer van der Pol on the other hand used it to a limited extent ([van der Pol 1929, p. 865] and [Niessen and van der Pol 1932, pp. 542–543]) and showed how it was transformed under the modified Laplace transformation:

$$p \int_0^{\infty} e^{-px} \delta(x)\, dx = p \tag{124}$$

or in van der Pol's notation $p \risingdotseq \delta(x)$. He also found the transforms of the derivatives of the δ-function and set down the list:

$$\begin{aligned} 1 &\risingdotseq [1], \\ p &\risingdotseq \delta(x), \\ p^2 &\risingdotseq \delta'(x), \\ p^3 &\risingdotseq \delta''(x), \end{aligned} \tag{125}$$

No actual definitions of the δ-function were given by van der Pol, but he described it as follows:

$\int_{-\infty}^{\infty} \delta(x)\, dx = 1$ and it is zero everywhere except at $x = 0$ where it is infinite.

Moreover, he gave the general property

$$\int_{-\infty}^{\infty} f(x)\delta^{(n)}(x)\, dx = (-1)^n f^{(n)}(0) \tag{126}$$

(see also [Koizumi 1931] for similar formulas).

The formulas (125) were of course needed only in those cases where δ-functions entered the original (electromagnetic) problem. Thus the Laplace transformation did not in itself stimulate the introduction of the δ-function. The Fourier transformation did however.

27. The Fourier transformation had been used for a long time to investigate periodic electromagnetic phenomena. In the 1920s it was adopted as another operational method for investigating transient electrical phenomena. Even in situations in which the original problem contained only ordinary functions f, the Fourier transforms $\mathscr{F}(f)$ could be distributions, for example,

$$\mathscr{F}(1) = \frac{1}{2\pi i}\,\delta. \tag{127}$$

For functions increasing at infinity more nonregular distributions were needed. In practical applications, therefore, there arose a need for a generalization, a need which was partially satisfied by the publication of tables of Fourier transforms making use of the δ-function. In 1928 G. A. Campbell published such a table with the explicit aim of helping the electrical engineers in their handling of transient phenomena. In his table one can find transformations like (127) or

$$\mathscr{F}((2\pi i f)^n) = \delta^{(n)}(g) \quad (f, g \text{ are the independent variables}) \tag{128}$$

and even more complicated expressions involving the δ-function (he termed the δ-function $\mathfrak{S}$, $\delta^{(n)}$ he called $\mathfrak{S}_n$). In his construction of the table he made strange use of the δ-function. His table was a

> table of paired coefficient functions. This means that if the coefficient $F(f)$ is employed with the cisoid $[e^{i2\pi f t}]$, and the coefficient $G(g)$ is employed with the unit impulse, and both products are summed for the entire infinite range of their parameters f and g, the same identical resulting time function is obtained.

That is, his pairs $F(f)$, $G(g)$ fulfil:

$$\int_{-\infty}^{\infty} F(f)e^{i2\pi f t}\,df = \int_{-\infty}^{\infty} G(g)\delta(t-g)\,dg. \tag{129}$$

This, of course, is only another way of saying $\mathscr{F}(F) = G$ since the right-hand side of (129) equals G. Campbell, however, preferred (129) because of the symmetry between F and G.[21] Campbell's table was extended and published as the book [Campbell and Foster 1931].

The fact that the Fourier transformation only applies to functions decreasing at infinity did not trouble only engineers but also mathematicians who from the middle of the 1920s gave generalizations of the Fourier transformation. These generalizations which avoided any explicit use of the δ-function, were discussed in Ch. 3.

28. The δ-function is often called "Dirac's δ-function" or just "the Dirac function" after the physicist P. A. M. Dirac (born 1902), for whom it was a forceful tool in his quantum mechanics. He had to use this improper function in order to draw an analogy between discrete and continuous variables which led him to a unified theory encompassing both matrix mechanics and wave mechanics. His theory was first published in [1926] in an article in the

Proceedings of the Royal Society of London and was developed further in his book *The Principles of Quantum Mechanics* [1930]. Dirac represented states of a mechanical system by vectors ψ [§7] and observables by linear operators [§9].[22,23] If, in the vector space, a finite or countably infinite basis ψ_p has been chosen (Dirac assumes this to be possible), the vectors ψ can be represented by their set of coordinates a_p with respect to the basis and operators α can be represented by finite or infinite matrices α_{pq}. In this case Dirac talked about a representation [§20]. In the representation above he thus found:

$$\psi = \sum_p a_p \psi_p, \tag{130}$$

$$\alpha\psi_q = \sum_p \psi_p \alpha_{pq}. \tag{131}$$

The physically interesting representations are those in which the basis consists of eigenvectors for an operator corresponding to some observable or of simultaneous eigenvectors for a set of commutable operators [§26]. In such representations however, the above countability requirement is usually not fulfilled since most physically important operators have continuous spectra [§22] and thus

> the total number of independent states [is] infinite and equal to the number of points on a line. The condition (130), which expresses that any ψ is a linear function of the fundamental ψs [the basis vectors], must now be rewritten with an integral instead of a sum, thus
>
> $$\psi = \int a_p \psi_p \, dp. \tag{132}$$

But this caused a problem:

> It is not strictly true that every ψ can be expressed in the forms (132) when the coefficients a_p are restricted to be finite, which is, of course, implied when one says they form a function of the continuous variable p. An example of a ψ that cannot be expressed in this form is one of the fundamental ψs, ψ_q say, itself. Another example is $\partial\psi_q/\partial q$. [Dirac 1930, §22.]

But having introduced δ [in §22] he could represent ψ_q by the coordinates:

$$a_p = \delta(p - q), \tag{133}$$

> where the improper function $\delta(x)$ is defined by
>
> $$\int_{-\infty}^{\infty} \delta(x)\, dx = 1, \tag{134a}$$
>
> $$\delta(x) = 0 \quad \text{for } x \neq 0. \tag{134b}$$

The derivative could be represented by

$$-\partial\psi_q/\partial q = \int_{-\infty}^{\infty} \psi_p \delta'(p - q)\, dp. \tag{135}$$

The representation of operators in the continuous case also required the δ-function. Instead of (131) Dirac wrote

$$\alpha\psi_q = \int \psi_p \alpha_{pq} dp \tag{136}$$

so that obviously the operator, multiplication by the constant c, was represented by the kernel $\alpha_{pq} = c\delta(p - q)$, which Dirac interpreted as the continuous analogue of the unit matrix δ_{pq}, δ being the Kronecker symbol. In this way he could adapt the matrix formulas to the continuous case by writing integrals instead of summations and $\delta(p - q)$ instead of δ_{pq}. This was surely the reason why he gave this improper function the name δ, not for "Dirac" but to parallel the Kronecker δ.[24]

29. Dirac gave "certain elementary properties of the δ-function which are deducible from, or at least not inconsistent with the definition". These are:

$$\delta(-x) = \delta(x), \tag{137}$$

$$x\delta(x) = 0, \tag{138}$$

$$\int_{-\infty}^{\infty} f(x)\delta(x)\,dx = f(0), \tag{139}$$

$$\int_{-\infty}^{\infty} f(x)\delta(x - a)\,dx = f(a), \tag{140}$$

$$\int_{-\infty}^{\infty} \delta(a - x)\delta(x - b)\,dx = \delta(a - b). \tag{141}$$

For the derivative which is "of course, an even more discontinuous and improper function than $\delta(x)$ itself", he found

$$\delta'(-x) = -\delta'(x), \tag{142}$$

$$x\delta'(x) = -\delta(x), \tag{143}$$

$$\int_{-\infty}^{\infty} f(x)\delta'(x - a)\,dx = -f'(a). \tag{144}$$

In later editions more formulas were added to this list. Thus in the third edition [1947] one finds [§15]:

$$\delta(ax) = a^{-1}\delta(x), \tag{145}$$

$$\delta(x^2 - a^2) = \tfrac{1}{2}a^{-1}\{\delta(x - a) + \delta(x + a)\} \qquad a > 0, \tag{146}$$

$$\frac{d}{dx}\log x = \frac{1}{x} - i\pi\delta(x), \tag{147}$$

and the division rule

$$A = B \Rightarrow A/x = B/x + c\delta(x). \tag{148}$$

All of this shows Dirac as a skillful manipulator of the δ-function. Some of the above theorems, especially (146), are not even obvious in distribution theory, since the changes of variables are hard to perform in $\mathscr{D}'$. Formula (136), indicating that all operators can be represented by a kernel, is an imprecise statement of what is today called the kernel theorem. Also the explicit representation of the identity operator by $\delta(p - q)$ anticipates this theorem which Schwartz proved in [1950]. He wrote in his autobiography [1974]:

> La physicien P. A. M. Dirac avait senti intuitivement le théorem des noyaux [kernel], disant que toute application linéaire continue de $\mathscr{D}_y$ dans $\mathscr{D}'_x$ peut s'exprimer, d'une manière unique, par un noyau-distribution, $K_{x,y}$, sous la forme
>
> $$T_x = \int K_{x,y}\varphi(y)\,dy \tag{149}$$
>
> il ne pouvait l'exprimer qu'en termes extrêmement vagues, c'est néanmoins lui qui m'en a donné l'idée et ma démonstration dans [1950], assez compliquée du point de vue de l'analyse fonctionelle telle qu'elle existait alors.

In the third edition of Dirac's *The Principles of Quantum Mechanics* he mentioned the "alternative way of defining the δ-function" as the derivative of the Heaviside function. He surely knew this definition and other properties of the δ-function from his student days; Dirac was educated as an electrical engineer.

30. Dirac's book on quantum mechanics became a most influential classic, published in four English editions and in a French and a German translation; with it the δ-function became a tool generally applied by physicists. In 1926, when Dirac's first article on the new treatment of quantum mechanics was published similar ideas were being explored by a group of mathematicians and physicists in Göttingen, including D. Hilbert, J. von Neumann and L. Nordheim. Inspired by Hilbert's work on integral equations, they assumed as Dirac did (136) that operators T associated with physical observables were integral operators

$$Tf(x) = \int \varphi(x, y)f(y)\,dy. \tag{150}$$

For this reason their quantum theory also depended heavily on the δ-function. In this way the δ-function came into the hands of mathematicians. However, during the same year that the joint work at Göttingen was published [Hilbert, von Neumann and Nordheim, 1927], one of its authors, von Neumann, dissociated himself from it and formulated a new and revolutionary basis for quantum mechanics [Von Neumann 1927]. His ideas were printed in book form in [1932] as *Mathematische Grundlagen der Quantenmechanik*, another highlight in the history of quantum mechanics. In his introductory remarks von Neumann proved that the representation of the identity operator by (150) would require that φ had the properties (134a) and (134b) which were incompatible with both the Riemann and the Lebesgue integrals.

> Dirac fingierte trotzdem die Existenz einer solchen Funktion

wrote von Neumann [1932, §3] and concluded:

> Wir wollen diese Gedankengänge die durch Dirac und Jordan zu einer einheitlichen Theorie der Quantenvorgänge ausgestattet wurden, hier nicht weiter verfolgen. Die "uneigentlichen" Gebilde (wie $\delta(x)$, $\delta'(x), \ldots$) spielen in ihnen eine entscheidende Rolle—sie liegen ausserbhalb des Rahmen der allgemein üblichen mathematischen Methoden, und wir wollen die Quantenmechanik mit Hilfe dieser letzteren beschreiben.

Von Neumann's idea was to describe quantum mechanics in an abstract (separable) Hilbert space, the axiomatic definition of which he set down for this purpose [1927, 1930]. He also represented observables as (unbounded) operators. By using the abstract operator theory in Hilbert spaces however, he avoided the difficulty with the integral representation (150). Another problem in the theory had been how to express a state ψ in terms of the basis of eigenfunctions of an operator (observable) (132). Not only may the coefficients a_p be generalized functions of p, but the eigenfunctions may lie outside the basic Hilbert spaces. This is, for example, the case with the momentum operator $\partial/\partial x_i$ whose "eigenfunctions" e^{ax_i} lie outside $L^2(\mathbb{R}^3)$ and even more the case with the position operator x_i whose "eigenfunctions" intuitively must be the functions $\delta(x_{i0} - x_i)$. Von Neumann overcame these difficulties by establishing the spectral theory for normal operators in Hilbert spaces in which eigenvalues in the continuous spectrum do not (in general) correspond to an eigenfunction and where the expression (132) is replaced by a Stieltjes like integral

$$\psi = \int_\sigma dE_\lambda, \tag{151}$$

where E_λ is a family of projections.

In this way von Neumann was able to escape the use of the δ-function, but many physicists continued to use Dirac's method. Thus the δ-function still played a great role in the further development of quantum mechanics.

31. In quantum field theory we find the most singular generalized function used prior to the rigorous theory of distributions. It was introduced in Jordan and Pauli's article, "Zur Quantendynamik ladungsfreiher Felder" [1928], under the name "der relativistisch invarianten Δ-Funktion", as the improper limit Δ of the sequence

$$\Delta_N(x, y, z, ct) = \iiint\limits_{|k|<N} \frac{2}{|k|} \sin 2\pi(k_x x + k_y y + k_z z - |k|ct)\, dk_x\, dk_y\, dk_z \tag{152}$$

$$(|k| = \sqrt{k_x^2 + k_y^2 + k_z^2})$$

for N tending to infinity. Let V_4 be a domain of integration in $\mathbb{R}^4$, V_3^+ its three-dimensional intersection with the positive light cone $r + ct = 0$ $(r = \sqrt{x^2 + y^2 + z^2})$ and V_3^- its intersection with the negative light cone $r - ct = 0$. Then they showed that

$$\begin{aligned}\int_{V_4} f(x, \dots, t)\Delta(x, \dots, t)\, dV_4 \\ = \int_{V_3^+} f(x, y, z, ct = -r)\frac{1}{r}\, dx\, dy\, dz \\ - \int_{V_3^-} f(x, y, z, ct = r)\frac{1}{r}\, dx\, dy\, dz.\end{aligned} \tag{153}$$

Und diese Gleichung ist jetzt als Definition der relativistisch invarianten.... Δ-Funktion anzusehen, unabhängig von ihrer speziellen Realisierung durch die folge (88).

A year later Heisenberg and Pauli showed that Δ could be expressed by the Dirac δ-function as

$$\Delta(x, y, z, t) = \frac{1}{r}(\delta(r + ct) - \delta(r - ct)) \tag{154}$$

["Zur Quantendynamik der Wellenfelder", 1929].

Pauli later found Δ important for the study of "The connection between spin and statistics" [1940] and stated explicitly that $\Delta/4\pi$ was a solution to

$$\begin{cases} \Delta u - \dfrac{\partial^2 u}{\partial t^2} - \kappa^2 u = 0, \\ u(\bar{x}, 0) = 0, \qquad \left(\dfrac{\partial u}{\partial t}\right)_{t=0} = \delta(\bar{x}). \end{cases} \tag{155}$$

for $\kappa = 0$; the solution to (155) for $\kappa \neq 0$ he found to be

$$-\frac{1}{4\pi r}\frac{\partial}{\partial r}F(r, t) \tag{156}$$

with

$$F(r, t) = \begin{cases} J_0(\kappa(t^2 - r^2)^{1/2}) & \text{for } t > r \\ 0 & \text{for } r > t > -r \\ -J_0(\kappa(t^2 - r^2)^{1/2}) & \text{for } -r > t \end{cases} \tag{157}$$

The function Δ is precisely the fundamental solution to the wave equation and agrees with Kirchhoff's singular function (28) for negative values of t.[25,26] The formulas above illustrate the extent to which difficult calculations with distributions were actually performed, and the great extent to which the δ-function entered into quantum mechanics.

32. A generalization of the three-dimensional δ-function and applications of it in electrostatics, quantum electrodynamics and nuclear physics was proposed by F. J. Belinfante [1946]. His starting point was the Fourier integral

$$\delta(\bar{x}) = \left(\frac{1}{2\pi}\right)^3 \int e^{i\bar{k}\bar{x}}\, d\bar{k} \tag{158}$$

for the three-dimensional δ-function (§8), from which he derived the two new tensor fields:

$$\delta_{ij}^{\text{lon}}(\bar{x}) = \left(\frac{1}{2\pi}\right)^3 \int \frac{k_i k_j}{k^2}\, e^{i\bar{k}\cdot\bar{x}}\, d\bar{k} \tag{159}$$

and

$$\delta_{ij}^{\text{tr}}(\bar{x}) = \delta_{ij}\delta(\bar{x}) - \delta_{ij}^{\text{lon}}(\bar{x}) \quad (\delta_{ij} \text{ is Kronecker's symbol}). \tag{160}$$

The importance of these new distributions is that they can separate the longitudinal part $A^{\parallel}$ and the transversal part $A^{\perp}$ of a vector field:

$$\begin{aligned} A^{\parallel}(\bar{x}) &= \int d\bar{x}'\, A(\bar{x}')\delta^{\text{lon}}(\bar{x} - \bar{x}'), \\ A^{\perp}(\bar{x}) &= \int d\bar{x}'\, A(\bar{x}')\delta^{\text{tr}}(\bar{x} - \bar{x}'), \end{aligned} \tag{161}$$

where $A^{\parallel}$ and $A^{\perp}$ are determined by

$$\begin{aligned} A(\bar{x}) &= A^{\parallel}(\bar{x}) + A^{\perp}(\bar{x}), \\ \text{rot}\, A^{\parallel}(\bar{x}) &= \text{div}\, A^{\perp}(\bar{x}) = 0, \\ A^{\parallel}(\infty) &= A^{\perp}(\infty) = 0. \end{aligned} \tag{162}$$

Still another application of the δ-function was made in statistical mechanics. We have already noted the implicit use of it in [Weber and Gans 1916] (note 14). A more direct application can be found in Casimir's paper "On Onsager's principle of microscopic reversability" [1945].

33. One can certainly add numerous other examples of applications of the δ-function and related generalized functions prior to the theory of distributions. From the examples given it should be clear, however, that the δ-function played a role in diverse physical theories and to a certain extent in mathematics, its main applications being in electrical engineering and quantum mechanics. Notwithstanding criticisms raised concerning the lack of a rigorous foundation the applications of the δ-function increased in frequency toward 1945. This trend was further reinforced by Schwartz' rigorization. By then, however, the δ-function was not only of help to physicists or nonrigorous mathematicians, but it became a powerful tool in pure mathematics, probably the most important single distribution.

34. Now let us turn from the applications to the mathematical foundation for the δ-function, first summarizing the different definitions used for it. In many cases it is impossible to distinguish between what the different authors considered as definitions and what they regarded as properties deducible from the definition. Most authors did not give an explicit definition, probably because they knew that any definition would be self-contradictory.

Four different definitions or characteristic properties were mentioned in the literature before 1945:

(a) $\delta(x) = (d/dx)H(x)$;
(b) $\delta(x) = \lim_{n\to\infty} f_n(x)$ (or $\delta = \sum_{n=0}^{\infty} f_n$) for suitable functions f_n;
(c) $\delta(x) = 0$ for $x \neq 0$, and $\int_{-\infty}^{\infty} \delta(x) = 1$;
(d) $\int_{-\infty}^{\infty} \delta(x-a) f(x)\,dx = f(a)$, or $\int_{-\infty}^{\infty} \delta(x) f(x)\,dx = f(0)$.

(a) Heaviside regarded this property as best characterizing the impulsive function; it was widely used later by electrical engineers. In the hands of Sumpner and Smith it was used, together with (b) and a loose idea about infinitely large and infinitely small quantities, in an attempt at a rigorous theory (see §38–44). This definition of the δ-function anticipates a definition of distributions as equivalence classes of pairs (D, f) of a differential operator and a function (see Appendix, §3).

(b) Many physicists point to this definition as the one which is most consistent with the physical intuition of a point mass or charge as the limiting case of very small particles. Kirchhoff used the description in this way in giving

$$F(\zeta) = \frac{\mu}{\sqrt{\pi}} e^{-\mu^2\zeta^2} \quad (\mu \text{ a very large positive constant}) \tag{163}$$

as an example of a δ-function (his fundamental definition was (c)). Similar examples were given by van der Pol [1929, p. 865] and Josephs [1946] (§25), and it is by such a sequence that Dirac brought his reader to definition (c). Limit definitions of a different type were considered in connection with Fourier series or Fourier integrals; in this case the f_ns were trigonometric functions. Such a description of δ can be found in [Fourier 1822] (§18 and 19), [Heaviside 1899] (§22) and [Jordan and Pauli 1928] (§31).

Definitions along these lines anticipated the definition of distributions as equivalence classes of certain series of functions (see Appendix, §2).

(c) Dirac (§28) made definition (c) the standard definition of the δ-function. It had already been noticed as a characteristic property by Kirchhoff [1882] (§6) and Heaviside [1899] (§21). It might be said to correspond to a definition of δ as a measure or as a Stieltjes integral, although this is not as obvious a rigorization as the two mentioned in (a) and (b).

(d) This is the definition which is closest to the definition of the δ-function as a functional. The property was used in all applications of the δ-function and was mentioned explicitly as a characteristic property by most authors, for example, Fourier [1822, §235, 3°] (quoted in §18), Heaviside [1899, §267] (105) and Dirac [1930, §22] (140). A variation on definition (d) was presented by Campbell [1928, p. 651] who stated after having approximated $\delta^{(n)}$ according to definition (b):

> It is necessary only that the method of approach to the limit give the same set of moments,

and more specifically he wrote:

> $\mathfrak{S}_n(g)[\delta^n(g)]$ is characterized by having all of its moments about the origin vanish except the nth moment, which is equal to $(-1)^n n!$

The mth moment of a function $f(x)$ in the interval $[a, b]$ is the integral:

$$\int_a^b f(x)x^m\,dx. \tag{164}$$

Thus Campbell characterized the distributions $T = \delta^n$ by their values

$$T(x^n).$$

In the distribution sense (i.e. in $\mathscr{E}'$) this uniquely determines T, since polynomials are dense in $\mathscr{E}$.[27]

Jordan and Pauli [1928] explicitly used a property similar to (d) to define the Δ-function (citation below (153)), and Heisenberg and Pauli [1929, pp. 11 and 12] used it to define δ and δ':

> Es ist zweckmäßig, dieses Resultat mittels des von Dirac eingeführten singulären Funktionssymbols $\delta(x)$ zu formulieren, das durch
>
> $$\int_a^b f(x)\delta(x)\,dx = \begin{cases} f(0) & \text{wenn } x = 0 \text{ in } (a, b), \\ 0 & \text{sonst} \end{cases}$$
>
> definiert ist.... Wenn wir dagegen die Ableitung der δ-Funktion in der üblichen Weise definieren, nämlich
>
> $$\int_a^b f(x)\delta'(x)\,dx = \begin{cases} -f'(0) & \text{wenn } x = 0 \text{ in } (a, b), \\ 0 & \text{sonst,} \end{cases}$$
>
> was aus (17) formal durch partielle Integration und Fortlassen der von den Grenzen herrührenden Glieder entsteht,...

These definitions are closer to Schwartz' ideas than any other interpretation of the δ-function.

Dirac also came very close to a functional interpretation in the third edition of his *The Principles of Quantum Mechanics* [1947]:

> Equations (139) and (140) show that, although an improper function does not itself have a well-defined value, when it occurs as a factor in an integrand the integral has a well-defined value. In quantum theory, whenever an improper function appears, it will be something which is to be used ultimately in an integrand. Therefore it should be possible to rewrite the theory in a form in which the improper functions appear all through only in integrands. One could then eliminate the improper functions altogether. The use of improper functions thus does not involve any lack of rigour

> in the theory, but is merely a convenient notation, enabling us to express in a concise form certain relations which we could, if necessary, rewrite in a form not involving improper functions, but only in a cumbersome way which would tend to obscure the argument.

35. Although the different "definitions" of the δ-function before Schwartz' definition anticipate different rigorous theories for it, they do not in themselves provide a mathematical foundation for operating with generalized functions, nor were they usually intended to give such a foundation. However, there are a few exceptions which I shall now investigate. Dirac's argument above is one such exception. He stressed explicitly that only with integrals do the improper functions have a meaning; but his aim was not to give a functional definition of the improper functions but to eliminate them altogether. In the first edition of his *The Principles of Quantum Mechanics* [1930] he expressed this in the following manner:

> The introduction of the δ-function into our analysis will not be in itself a source of lack of rigour in the theory, since any equation involving the δ-function can be transcribed into an equivalent but usually more cumbersome form in which the δ-function does not appear.

This argument does not tell how the transcription should be done, but its presentation just following Dirac's introduction of δ as an improper limit of ordinary functions suggests that the rigorous method would consist in calculating with the approximations of the δ-function and then taking the limit in the result. Whether this was Dirac's idea or whether he thought of introducing suitable integrals, in which δ could be the integrand, it is clear from both the cited editions of his quantum mechanics that to establish rigor he felt it was sufficient that the same result be obtainable by a different method in a rigorous way. This kind of argument is identical with the one used to defend the integral methods before Newton and Leibniz; at that time the rigorous substitution which was assumed mostly without verification was the method of exhaustion.

36. A justification of the δ-function along these lines can be found in other places. Fourier [1822, §417] thus argued in the following way for the use of an infinite p in (95), i.e. for the use of the δ-function instead of its approximations:

> La démonstration précédent suppose la notion des quantités infinies, telle qu'elle a toujours été admise par les géomètres. Il serait facile de présenter le même démonstration sous une autre forme an examinant les changements qui résultent de l'accroissement continuel du facteur p sous le signe sin $p(\alpha - x)$. Ces considérations sont trop connues pour qu'il soit nécessaire de les rappeler.

Fourier's last remark clearly conceals his loose argument: a proof based on convergence for p tending to infinity was in no way well known to Fourier's contemporaries; it was first presented by Dirichlet several years later [1829] and [1837].

37. There are three ways to handle the argument mentioned above concerning substitution of a δ-proof with a rigorous method.

(a) First there is the rhetorical stance represented by Fourier and Dirac, who only postulated a possible alternative argument without giving it.

(b) The second way of pursuing the argument is to give both the loose δ argument and the more rigorous argument without the δ-function. This was the most infrequently applied way of handling the problem. Courant–Hilbert [1924] used it in the treatment of the fundamental solution (see §2). Casimir [1945], after obtaining a certain formula applying Fourier transforms, stated:

> This equation can be obtained somewhat more directly though less rigorously by using the ΔT instead of their Fourier components and making a liberal use of δ-functions.

Following this he carried out the nonrigorous computation.

(c) The most frequent method, used by mathematicians to treat phenomena which for their intuitive physical description involved δ-functions was to give only the rigorous method without any reference to the intuitive argument (based on the δ-functions). There are numerous examples of such treatment, the most explicit being the theory of Green's function—or fundamental solution and the theory of point, line and surface mass distributions in potential theory.[28] To view such theories as rigorizations of the δ-function is of course questionable since it is not always clear to what extent the particular mathematicians thought intuitively in terms of δ-functions. However, there is much evidence that these kinds of intuitive ideas were in the air at the time. For instance, Belinfante [1946] wrote that it was *well known* that

$$\Delta\left(\frac{1}{r}\right) = -4\pi\delta(\bar{x}) \quad \text{with } r = |x|. \tag{165}$$

This formula was sometimes given in a rigorous garb in connection with Green's theorem. Courant and Hilbert's analysis (§2) gives an explicit formulation of this intuitive idea. Also in connection with a discussion of "impulses", Wiener describes as "familiar" the idea of using Stieltjes integrals in order to rigorize the point, line and surface distributions in potential theory (citation Ch. 3, §7).

From this evidence I infer that at least many, if not all, mathematicians knew that such theories were a way to rigorize *arguments* involving δ-functions, but nobody thought of them as methods for rigorizing the δ-function itself. Had that occurred to anyone before 1945, the development of the theory of distributions would have progressed quite differently. For instance, it would have been very natural in connection with potential theory to define δ as a measure and with that as a starting point to develop a more general theory which would include its derivatives. Of course, it is not very fruitful to imagine what might have happened, but it is interesting to try to analyze why certain ideas, which we find natural today, were not pursued. In this case the answer is probably that although mathematicians intuitively felt the connection between their rigorous theories and the δ-function, they could not see the aggregate of problems involving generalized functions as a unified whole, as we can with the unifying theory of distributions. Thus mathematicians working with potential theory could not see that their method, if used to define the δ-function, could help physicists in quantum mechanics or electrical engineers using Heaviside's calculus.

This explanation, however, cannot account for the fact that von Neumann, who saw that the use of spectral measures could eliminate the δ-function from quantum mechanics, did not give a measure definition of δ. One reason for this is that the question of the derivatives would still have remained unsolved. Another reason could be that mathematicians at that time were so convinced about the illegitimacy of the δ-function that they restrained themselves from any attempts to rigorize it.

38. The rigorization methods analyzed in the previous section provided means for placing comprehensive arguments involving δ-functions on a firm basis, but they did not show how the δ-function itself could be made a respectable mathematical object. Before the theory of distributions only a few such attempts were given. The first of these was due to J. J. Smith from the General Electric Company, Schenectady, New York, who in a long article in three parts, "An analogy between pure mathematics and the operational mathematics of Heaviside by means of the theory of H-functions" [*Journal of the Franklin Institute*, 1925], tried to provide a suitable description not only of the δ-function but of the entire Heaviside calculus:

> In this paper an attempt is made to reconcile the operational methods of Heaviside with the methods of pure mathematics. In order to do so certain modifications in the conceptions of Function, Limit, Differential Coefficient, etc., are suggested. The new definitions are called H-functions, etc., and have the property that they give the same results as the present definitions at ordinary points, i.e. points at which there are no discontinuities, etc. The difference in the results obtained by both methods is pointed out by reference to well-known cases.

This could be a general description of the theory of distributions, but Smith's "reconciliation" is very different from this theory. It was based on the standard of operational mathematics and did not fulfil the requirements for rigour in ordinary mathematics. For example, he found it "best, although perhaps not strictly logical, to develop in an experimental way the theory which follows".

39. In accordance with his experimental idea Smith defined the more complex ideas such as H-differentiation before he defined the basic concept of H-function. The H-derivative of an H-function f in $[a, b]$ (which may be multiple-valued) was defined as the H-limit of

$$\frac{f(x + \Delta x_1) - f(x - \Delta x_2)}{\Delta x_1 + \Delta x_2}, \tag{166}$$

when $\Delta x_1, \Delta x_2 \to 0$ and $\Delta x_1/\Delta x_2 = p$, with p an arbitrary but fixed constant, $0 \leq p \leq \infty$. All the values obtained for different ps were values of the H-derivative. As a most important example [pp. 523–527] he found the successive derivatives of the H-function $y(x)$ defined by:

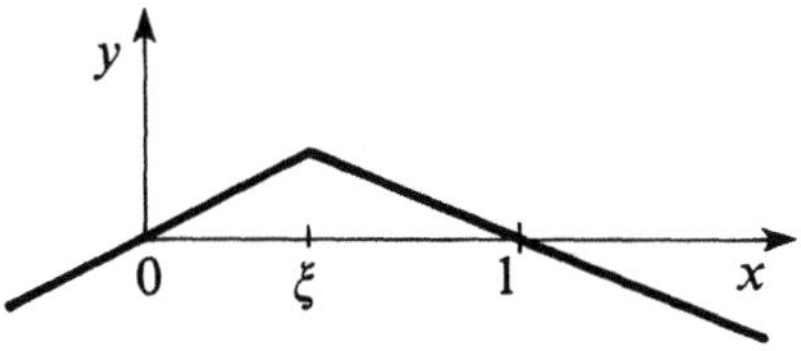

$$f: \begin{cases} y = (1-\xi)x, & x \leq \xi, \\ y = (1-x)\xi, & x \geq \xi. \end{cases} \tag{167}$$

The first H-derivative of (167) he found to be

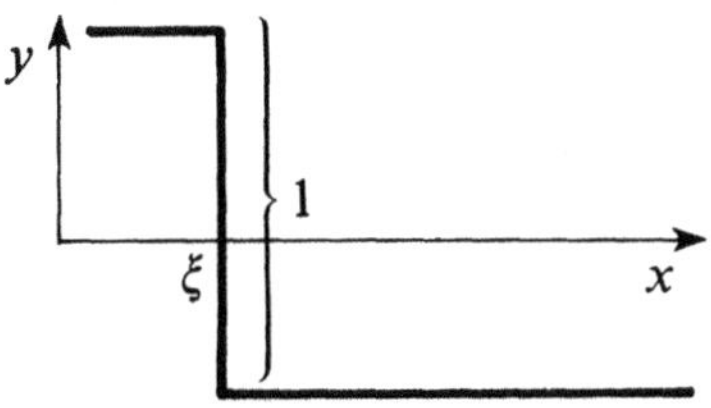

$$f': \begin{cases} y = 1 - \xi, & x < \xi, \\ y = -\xi, & x > \xi, \\ y = \dfrac{1}{1+p} - \xi, & x = \xi \\ (0 \leq p \leq \infty) \end{cases} \tag{168}$$

or in ordinary language $f' = -H(x-\xi) + 1 - \xi$. For the second derivative he obtained the expression:

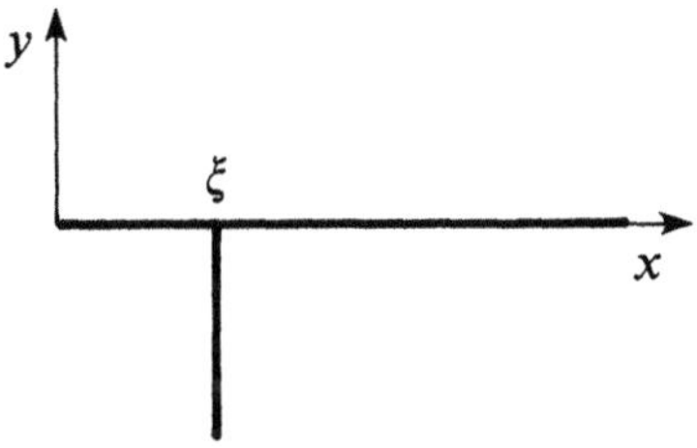

$$f'': \begin{cases} y = 0, & x < \xi, \\ y = 0, & x > \xi, \\ y = -k, & x = \xi \quad (0 \leq k \leq \infty) \end{cases} \tag{169}$$

or $f'' = -\delta_\xi$. All of the successive derivatives took the form

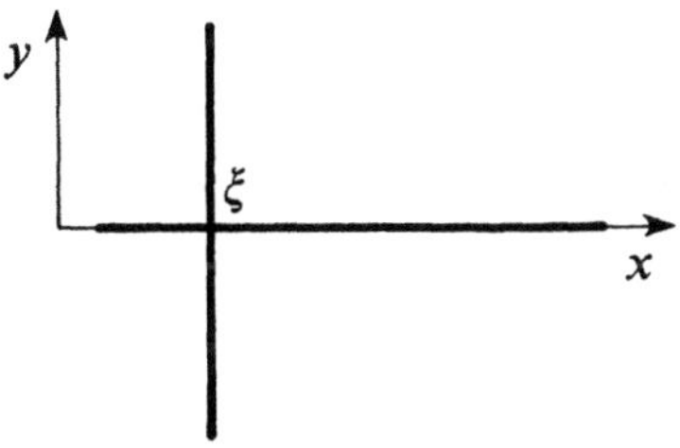

$$f^{(n)}\ (n \geq 3): \begin{cases} y = 0, & x < \xi, \\ y = 0, & x > \xi, \\ y = -k, & x = \xi \\ (-\infty \leq k \leq \infty). \end{cases} \tag{170}$$

> From these last two statements [i.e. (170)], it would appear as if the third and fourth H-derivatives of (167) were the same H_p-function... it is only due to the insufficiency in the definition that they appear the same.

Smith first tried to "save the phenomena" by introducing different orders of infinity, $\infty^1, \infty^2, \ldots$, but he found [p. 530] that this was "perhaps not the best way to proceed in defining the various H-derivatives" and proposed that they were best called the first, second, third, etc. H-derivatives of (167). "This", he claimed [p. 657], "makes their definition quite precise."[29]

40. The δ-functions are examples of H-functions, which Smith defined in the following way [p. 656]:

> An H-function will now be defined as follows: Let x lie in the continuous interval $a \leq x \leq b$. Then if a formula or norm is given for determining the value of y in the neighborhood of every point x in this interval, y is said to be an H-function of x. In case that the value of y tends to more than one value in the neighborhood of any point x, a definite limiting process must be given to determine fully the characteristic properties of the H-function at this point. The term "in the neighborhood of" is to be understood to include the point itself as one point in its own neighborhood.

This definition is so obscure that it is difficult to have any idea of what an H-function is supposed to be. The examples mentioned by Smith may lessen the obscurity. An H-function would typically be given as an H-sum or an H-limit of nice analytic expressions such as the H-sum:

$$y = \sin x \sin \alpha + \sin 2x \sin 2\alpha + \cdots \tag{171}$$

describing the δ-function [p. 652]. The limit of the curve:

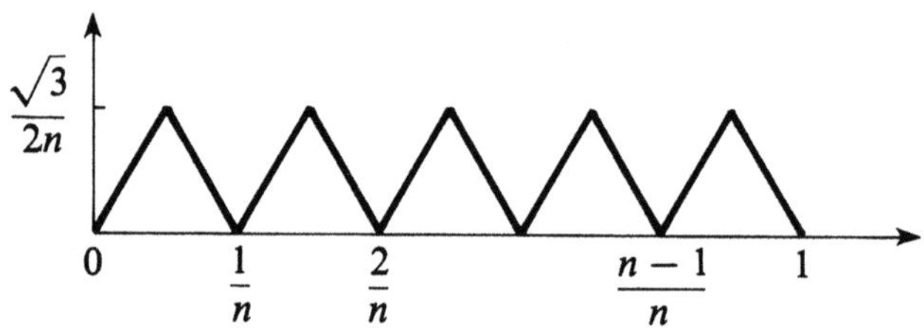

Figure 7

for n tending to infinity is also an example of an H-function.

The H-limit was characterized in this way [p. 659]:

> Thus the H-limit may be defined in general as the aggregate of the limits of all the characteristic properties of the H-function, where, for the present, these limits may be taken in the ordinary manner. It might appear to be not a very easy matter to define what "all the characteristic properties of an H-function" are.

This does not in fact become clear in Smith's article. But one is given a vague idea about the kinds of characteristic properties he considered by looking at Figure 7. Smith argued that since the length of the saw-toothed curves was 2 for all n, then the H-limit had length 2, whereas in real function theory it would have length 1 [p. 649] (cf. Young's generalized curves, Ch. 2, §26). He thought of the limit of an oscillatory curve as oscillating, a fact which can also be seen in his treatment of the H-limit of

$$y_n = \frac{\sin (2n + 1)\beta}{\sin \beta} = \frac{1}{2} + \cos 2\beta + \cos 4\beta + \cdots, \tag{172}$$

which he conceived of as "a curve filling the whole of the space between the boundaries cosec β and $-$cosec β". However, he knew its δ-character very well, such as it had already been used by Fourier.

H-continuity which Smith did not define explicitly, was another characteristic property which was taken over in the limit. Therefore, "It is difficult to see how, if such a rule is adopted in framing or extending definitions, discontinuous H-functions can ever arise." For example, the δ-function and its derivatives became H-continuous when they were constructed as successive H-limits (H-derivatives) of a continuous function (167).

Another characteristic of H-limits was that "no limiting process had precedence over another" [p. 635]. Thus, for example, in the summation of a series of functions, which Smith wrote in the form

$$\lim_{k\to\infty} \lim_{x\to x_0} \sum_{n=1}^{k} f_k(x), \tag{173}$$

the n and x limits should be taken simultaneously. In the case of nonuniform convergence this gave rise to a multiple-valued sum.[30] Smith commented [p. 655]:

> It may be remarked that the H-limit of the sum of a series, if we consider its arithmetical properties alone, is the aggregate of the aggregate of points of accumulation of the sum in the neighborhood of the various values of the independent variable in any given interval.

41. There should be no need to point out that this obscure "reconciliation" of operational mathematics, including improper functions, with ordinary mathematics, is completely unsatisfactory. I am sure that all mathematicians who heard Smith's short report on these ideas at the International Mathematical Congress in 1928 agreed when Smith said in his concluding remarks:

> In general, the impression seems to be that the introduction of multivalued functions would lead to confusion and serve no useful purpose. [Smith 1928.]

The usefulness itself of Smith's theory of H-functions is highly doubtful. Electrical engineers knew perfectly well how to operate intuitively with the Heaviside calculus. The muddled theory of H-functions could not, as far as I can see, help them judge the correctness of their results. I have found no reference to Smith's theory in engineering texts. Even Smith's own operational calculations of certain Green's functions—characterized with the aid of the δ-function as in (1)—did not use the theory of H-functions in any crucial way.

It is very interesting to compare the theory of H-functions with the eighteenth- and early nineteenth-century ideas concerning the basic concepts in analysis, with which it has many points in common (see [Lützen 1978]), such as the function concept and to a certain extent the limit concept. There are differences as well, for example the definition of continuity which in the theory of H-functions seems to be a property of all H-functions. Moreover

Smith had the advantage over his eighteenth- and nineteenth-century predecessors that he could compare his theory with the rigorous classical definitions from the end of the nineteenth century and thus make clear what his theory was *not* supposed to be. However, such a comparison with Weierstrassian analysis would make it quite evident to mathematicians of the twentieth century that Smith had not succeeded in describing in a rigorous way what his theory *was*.

42. Although the theory of H-functions in many ways merely looks like an attempt to revive an antiquated style of mathematics, there are a few ideas which point to future developments. The idea of generalized functions without specified values in specified points proved to be a fruitful idea in the theory of distributions. In one place Smith was close to discovering what kind of meaning such multiple-valued functions could be given: having described the limit of y_n in (172) as a curve filling out a whole area, he wrote that such a function

> from a purely arithmetical point of view is meaningless. Looked at from the point of view of its integral between any two points in the interval, it has a definite meaning....

(hinting at (88)). He never carried this "functional" idea any further. The free interchangeability of different limit procedures was also paralleled in the theory of distributions, for example, by the continuity of differentiation, the importance of which was repeatedly stressed by Schwartz:

> Au §5 [1950/51, Vol. I, Ch. 3] on démontre une propriété fondamentale (p. 80, Théoréme XVIII), la continuité de la dérivation, qui permet la dérivation terme à terme ou sous le signe ∫ des suites, séries, intégrales convergentes; c'est là, avec la possibilité de dériver indéfiniment, l'avantage essentiel des distributions. [Schwartz 1950/51, Vol. I, p. 66.]

Moreover, Schwartz in his autobiography [1974, p. 5] wrote:

> Trouver une *théorie* qui rendait toutes les fonctions indéfiniment dérivables et permettait la dérivation terme à terme des séries convergentes, c'était exactement le genre de recherche qui me convenait.

In spite of these few hints of fruitful ideas the theory of H-functions was an unsuccessful first attempt to generalize the concept of function and the calculus so as to incorporate the δ-function and its derivatives. The theory does not seem to have had any influence on the development in which Smith's program was finally carried out more successfully.

43. Two of the components in the theory of H-functions were the infinitely large numbers, ∞^1, ∞^2, ∞^3, etc. and the infinitely often oscillating functions These two notions were central to another attempt at a rigorization of the δ-function put forward by the English mathematician E. W. Sumpner in 1931. However, whereas Smith's treatment of the infinitely large and infinitely small quantities had been very rough, Sumpner had a very refined way of handling them. His definition of the Heaviside function and the δ-function rested on these carefully chosen quantities:

> Let c be a positive infinitesimal which may be as small as we please, but which, once chosen, must be treated throughout analysis as a fixed quantity.
>
> Let n be a positive integer, chosen after c has been fixed, and so large that nc^2 is infinite.
>
> Let w be a positive infinity but one of lower order than $1/c$, and such that cw is infinitesimal.

With these nonstandard numbers at his disposal, he defined the Heaviside function, or the "unit" as

$$H(t) \equiv \frac{t^c}{c!} H_{01}(t) \equiv \frac{t^c}{c!}\left[\frac{1}{2} + \frac{1}{\pi}\,\mathrm{Si}(nct)\right], \tag{174}$$

where

$$\mathrm{Si}(x) = \int_0^x \frac{\sin u}{u}\,du \tag{175}$$

and the impulse function:

$$pH(t) \equiv \frac{d}{dt} H(t). \tag{176}$$

This definition of $\delta = pH$ was inspired by "the impulse aspect of Fourier's theorem" and if one forgets about the factor $t^c/c!$, it is in full accord with Fourier's (95). Sumpner proved that $H(t)$ for all real values in $[-w, w]$ had the usual properties of the Heaviside function, for example:

> For every value of t from $+c$ to $+w$, H will be unity except for a quantity which is highly infinitesimal compared with c.

Because of the factor $t^c/c!$ the corresponding statement for $t < 0$ is stronger:

> For every negative value of t from $-w$ up to and including absolute zero H will be zero except for a quantity which is highly infinitesimal compared with c.

Examined under a microscope, however, $H(t)$ is a "wave function of constant [infinitely small] wavelength, which fluctuates about two level values, these being unity for positive and zero for negative values of t". Moreover, looked at in this way, $H(t)$ is, according to Sumpner, so smooth that its derivatives can be taken and will be "continuous". "Its derivative pH is a wave function of the same wavelength, fluctuating about zero for all values of t." Here the last factor $H_{01}(t)$ which Sumpner investigates in [§4] is responsible for the oscillatory character of the function and $t^c/c!$ [§5] makes $H(t)$ vanish for all negative values of t. Although Sumpner thought that "the function $t^c/c!$, where c is a high infinitesimal, looks simple enough", he had to postulate that it was zero for $t < 0$, even if "this convention is not justifiable in itself".

44. Except for the possible exclusion of the last-mentioned point, Sumpner worked very carefully through the proof of the above properties of $H(t)$ and $pH(t)$ keeping accurate account of the different infinitely small and large numbers involved. In spite of this, Sumpner's treatment could not be considered a rigorous introduction of the δ-function, first because no proof existed that infinitely small and large numbers could be defined so that they, together with the ordinary real numbers, could be handled according to the ordinary rules of algebra, such as Sumpner had done. Secondly, Sumpner had not defined a differential operator p that could differentiate functions like H, which was smooth only "when looked at under a microscope".

The ideas of Smith and especially of Sumpner pointed to a solution to the problem of the δ-function in terms of infinitely small and infinitely large magnitudes. However, theories of this kind were not at hand before the work of Laugwitz [Laugwitz and Schmieden 1958] and Robinson [1961], who both quickly adapted their respective theories to a treatment of the theory of distributions: Laugwitz [1959, 1961], Robinson [1966]. Seen in the light of these modern theories Sumpner's analysis may be essentially right, but it certainly does not give the simplest representative of the δ-distribution.

45. Another potentially more promising theory of generalized functions emerged from the problems concerning a mathematical basis for the δ-function. The inventor of this second theory of generalized functions (Sobolev's theory was the first) was the Dutch physicist, H. A. Tolhoek. He wrote to me in a letter:

> I started to study physics in 1940 during the war. The conditions for studying during the war were bad in Holland. However, I studied some quantum mechanics from Dirac's book. I asked myself the question whether one could not make a decent mathematical theory for the Dirac δ-function and similar improper functions by considering them as "improper limits" of ordinary functions. I wrote up something about this in 1944 and sent this to Prof. L. Rosenfeld (then in Utrecht), who encouraged me to continue.

In this manuscript of approximately 15 pages an improper function

> is defined by a sequence of functions $f_n(x)$, e.g. $(n/\sqrt{\pi})e^{-n^2x^2}$ for $\delta(x)$ which does not converge in the ordinary sense, but for which we require that:
>
> $$\lim_{n\to\infty}\int f_n(x)f(x)\,dx \tag{177}$$
>
> exists for every function $f(x)$ that is differentiable a sufficient number of times. In this manner it is ... possible to give a justification of the main properties of improper functions. [Tolhoek 1949 on his 1944 paper.]

Two sequences f_n, g_n defined the same improper function if

$$\lim_{n\to\infty}\int f_n(x)f(x) = \lim_{n\to\infty}\int g_n(x)f(x) \tag{178}$$

for all functions f which are sufficiently smooth [Tolhoek 1978, Interview]. Tolhoek developed this small theory of generalized functions knowing only Dirac's work and von Neumann's criticisms of it. He was unaware of the work done in the theory of differential equations (Sobolev) or in the theory of Fourier transforms or in any of the other branches of mathematics in which distribution-like ideas were developed [Tolhoek 1978, Interview].

46. In spite of Rosenfeld's encouragement, Tolhoek set aside the generalized functions for some years in order to obtain his "doctorandus-degree" (similar to a master's degree) in physics and did not return to the generalized functions until 1948/49 when he wrote a manuscript of 122 pages on "The mathematical justification of the use of the Dirac delta-function and other improper functions with applications" [Tolhoek 1949].

> I had then learnt of some papers of L. Schwartz about the subject and I took them into account to a certain extent. My manuscript was partially a review, partially original work. [Letter, Oct. 1977.]

The definition of an improper function suggested in the 1949 manuscript differed from both the limit approach in the 1944 manuscript and from Schwartz' definition. Tolhoek defined it as follows:

> Suppose we have the differential operators Ω_k and the proper functions $F_k(x_1, \ldots, x_n)$ defined and absolutely integrable in a regular region R of $\mathbb{R}_n$ $(k = 1, 2, \ldots, N)$. We call the symbol
>
> $$D(x_1, \ldots, x_n) \equiv \sum_{k=1}^{N} \Omega_k F_k(x_1, \ldots, x_n) \tag{179}$$
>
> an improper function, if it is possible for every function $f(x_1, \ldots, x_n)$ which is defined and continuously differentiable a sufficient number of times in R, to transform by formal partial integrations:
>
> $$\int_R \left[\sum_k \Omega_k F_k(x_1, \ldots, x_n)\right] f(x_1, \ldots, x_n)\, dx_1, \ldots, dx_n \tag{180}$$
>
> into an integral, which exists in the ordinary sense. We define the *integral with the improper function* $D(x_1, \ldots, x_n)$:
>
> $$\int_R D(x_1, \ldots, x_n) f(x_1, \ldots, x_n)\, dx_1, \ldots, dx_n \tag{181}$$
>
> as this value. [Tolhoek 1949, p. 11.]

Two generalized functions were called equal if their integrals (181) were equal for all suitably smooth functions f.

47. Tolhoek defined several operations such as addition, subtraction and multiplication,[31] differentiation, convergence of *ordinary* functions to improper functions (improper limits), and Fourier and Laplace transforms of improper functions. In 1949 Bochner's work on generalized Fourier integrals was known to Tolhoek who proved that for a function f in F_k his Fourier transform was equal to

$$\frac{d}{ds^k} E(s, k), \tag{182}$$

where $E(s, k)$ is Bochner's k transform of f.

As examples of improper functions Tolhoek mentioned the δ-function and its derivatives and the Cauchy principal value, whereas Hadamard's partie finie was unknown to him. He applied the theory to different mathematical and physical problems such as finding Green's functions for various (partial) differential equations (Green's function $K(x, \xi)$ was defined as the solution of $L(K(x, \xi)) = \delta(x - \xi)$ where L is the differential operator). Among other things he found Green's function (92), (93) for equation (91). Moreover, he applied the improper functions to the theory of fluctuations (as Casimir had done [1945]) to quantum mechanics and to the study of vector fields, where he, among other things, proved $\Delta(1/r) = -4\pi\delta(\bar{x})$. All of Tolhoek's results were known to at least some extent, but his paper was the first place in which all the different aspects of the δ-function and more general improper functions were studied conjointly with proofs based on nearly rigorous ideas—Schwartz' short papers published before 1949 contained only few proofs. However, Tolhoek's second paper, like his first, was never published. He had sent it to the *Proceedings of the Dutch Academy*, but they found it too long and therefore refused it [Tolhoek 1978, Interview].

48. Even if it had been published, and thus had become the first comprehensive work on generalized functions, it would probably have had difficulty competing with Schwartz' theory of distributions. Tolhoek wrote to me that he worked

> rather from the point of view of a physicist. I did not use explicitly functional analysis and cannot claim the same mathematical rigour as obtained in the later books by L. Schwartz.

In the main the topological aspects of the improper functions were insufficiently treated by Tolhoek, but as a whole his theory met a much higher standard of rigour than any of the earlier theories of the δ-function, except once again for Schwartz' distributions. On the other hand, Tolhoek felt that his treatment and notation were more closely connected with physicists' ideas and easier to apply than Schwartz' theory:

> For the applications the introduction of new variables in improper functions is very important and in this connection it seems desirable to write improper functions as functions with an argument, e.g. $\delta(x)$, $\delta(x - a)$, etc., and not as functionals as Schwartz does. [Tolhoek 1949, p. 119.]

As it happened Tolhoek did not influence the development of the theory of distributions. He might have had an influence on Korevaar, who was his roommate during their student days in Utrecht. However when Korevaar in [1955] gave his interpretation of the distribution theory, which is very close to the one used in Tolhoek's first paper, he was able to refer to several other mathematicians who had arrived at the idea of defining distributions as improper limits, for example, Mikusiński [1948] and Temple [1953]. Another mathematician H. Könich gave in [1953] a definition of distributions very similar to Tolhoek's second definition, but his ideas were independent of Tolhoek's. Tolhoek concludes his letter to me:

> After all it may be rather a pity that I did not get anything published about 1949 concerning distribution theory. As it is, I am not making any claims.

49. It is striking that few attempts were made before 1945 to rigorize the δ-function compared with the many approaches available. I have already discussed some of the reasons for this in §37. Another reason may be that physicists and engineers did not bother too much about the foundations when they knew that the results of the calculations were correct. This attitude is completely understandable as long as practical computations only, as in electrical engineering, depended upon the δ-function. However, when theoretical insight into the basic structure of nature came to depend on this illegitimate function, as it happened in quantum mechanics, criticism became more severe. Another factor that made the use of the δ-function in quantum mechanics more criticized than its use in electrical engineering had ever been was the interest that leading mathematicians as, for example, von Neumann, took in this field.

When Schwartz developed his rigorization of the δ-function, most physicists were relieved to see that their methods could be made exact, but in practice most of them continued to work with δ-distributions as if the theory of distributions had never been created. There even were physicists who apparently did not know of the theory of distributions several years after its publication: for example, in [1957], Infield and Plebański wrote a note, "On a further modification of Dirac's δ-functions", in which they loosely gave a continuous approximation $\delta_\varepsilon(\vec{x})$ for the three-dimensional δ-function. Even in mathematical texts after 1950 one may find δ-functions introduced in a nonrigorous way: for example, in the second edition of van der Pol and Bremmer's book on operational calculus [1955], one finds only a very short note on the theory of distributions; otherwise the theory of the δ-function is treated in a most intuitive and poorly established way.

The δ-function together with its derivatives was undoubtedly the most significant generalized function prior to the theory of distributions. Although it was not the direct starting point for Schwartz' discovery of the theory, it played a key role in his further work (see Ch. 6).

Chapter 5

De Rham's Currents

1. Naturally enough, the prehistory of the theory of distributions was concentrated within analysis and mathematical physics. However, there was one exception: de Rham's currents which will be discussed in this chapter.[1]

Schwartz wrote in his historical introduction [1950/51]:

> Enfin il est un domaine tout différent où les distributions jouent aussi un rôle. En topologie algébrique, l'homologie d'une variété différentiable est donnée soit par les "chaînes singulières", soit par les formes différentielles, avec d'un côté l'opération "bord", de l'autre l'opération "différentielle extérieure". D'où l'idée naturelle de faire une synthèse entre ces deux catégories d'êtres. C'est M. de Rham, qui eut l'idée d'introduire les "courants", comprenant á la fois comme cas particuliers les chaînes et les formes, et une opération de dérivation qui était (au signe près éventuellement) le bord pour une chaîne et la différentielle extérieure pour une forme. La théorie des courants est simplifiée et perfectionnée par celle des distributions-formes différentielles sur une variété, qui englobe aussi les résultats de P. Gillis [1943], cette théorie sera publiée dans un article à part.

2. De Rham (born 1903) fused the ideas of chains and differential forms in [1936]. As remarked by Schwartz in the quotation above De Rham saw that associated with each notion was an operation, the boundary operation b and the exterior derivative d, respectively, with the properties:

$$bbc = 0 \qquad \text{for all chains } c, \tag{1}$$

$$dd\omega = 0 \qquad \text{for all differential forms } \omega, \tag{2}$$

(for an explanation of the algebraic topological terms see note 2).

The connection between the operators d and b is even more striking when seen in the light of Stokes' theorem

$$\int_c d\omega = \int_{bc} \omega,$$

For chains, Poincaré had already introduced the concept of homology at the end of the nineteenth century. A cycle was said to be homologous to zero if it were the boundary of a chain. Two cycles were said to be homologous, if their difference were homologous to zero. Inspired by this definition E. Cartan [1928, 1929] and de Rham [1929], introduced homology for closed forms: a form is homologous to zero if it is the differential of a form. Two forms are said to be homologous if their difference is homologous to zero.

De Rham [1936] could now prove the two theorems:

Theorem I. *The p-cycle c is homologous to zero and if and only if*

$$\int_c \omega = 0 \quad \textit{for all closed 'p forms } \omega. \tag{3}$$

Theorem II. *The closed p-form ω is homologous to zero if and only if*

$$\int_c \omega = 0 \quad \textit{for all p-cycles } c. \tag{4}$$

Two similar conditions for the zero homology are given in Poincaré's duality theorem:

Poincaré's theorem. *A p-dimensional cycle c^p is homologous to zero if and only if*

$$I(c^p \cdot c^{n-p}) = 0 \tag{5}$$

for all $(n - p)$-dimensional cycles c^{n-p}. Here $I(c^p \cdot c^{n-p})$ denotes the algebraic number of intersections between c^p and c^{n-p},

and the following:

Theorem III. *The p-form ω_1 is homologous to zero if and only if*

$$\int_V \omega_1 \omega_2 = 0 \tag{6}$$

for all closed $(n - p)$-forms ω_2.

De Rham gave the sketch for a derivation of Theorem I from Poincaré's well-known theorem. His proof was indirect, i.e. he proved that for each p-cycle c which was not homologous to zero there existed a p-form ω such that

$$\int_c \omega \neq 0. \tag{7}$$

He considered a three-dimensional manifold in $\mathbb{R}^3$ and a two-dimensional cycle c^2 which was not homologous to zero. According to Poincaré's duality theorem there exists a 1-cycle c^1 such that

$$I(c^1 \cdot c^2) \neq 0. \tag{8}$$

c^1 consists of closed oriented curves. De Rham thought of these as cables carrying an electrical current of unit intensity. The total current through the surface c^2 is equal to $I(c^2 \cdot c^1)$, i.e. it is different from zero.

On conçoit ensuite la possibilité d'étaler un peu ce courant, de manière qu'il remplisse une sorte de tube entourant c^1, avec une intensité de volume contenue à l'intérieur du tube et nulle sur sa frontière. La forme ω, égale au debit élémentaire de ce courant dans le tube et nulle en dehors satisfait à toutes les conditions [i.e. (7)] requises. [De Rham 1936, p. 220.]

De Rham's outline for a proof is interesting from the point of view of distribution theory since the idea of approximating the current in a curve by a current in space is closely related to the approximation of the δ-distribution by smooth functions. Moreover, it shows that a current can be represented both by a one-dimensional cycle and a form of degree 2. Such a correlation between p-forms and $(n - p)$-cycles is not only seen from the proof, but also from a comparison of Theorem I and Poincaré's theorem and of Theorem II and Theorem III.

Cela suggère l'idée que dans une variété à n dimensions V, un p-champ et une $(n - p)$-forme doive être deux aspects d'une même notation plus générale, que j'appellerai courants à p dimensions. [De Rham 1936, pp. 220–221.]

3. Thus, inspired by the similarities between forms and chains, de Rham in [1936] defined the concept of a current. He defined an elementary current as a pair (c^{p+k}, ω^k) consisting of a $(p + k)$-chain and a k-form; a current is then a linear combination of elementary currents. He identified the current (V, ω) with the form ω, and the current $(c, 1)$ with the chain c. For the set of currents de Rham defined the operations of addition and multiplication, both multiplication by scalars and multiplications of two currents. Moreover, he defined the differential of a p-current (c^{p+k}, ω^k) to be the following $(p - 1)$-current

$$d(c, \omega) = (c, d\omega) + (-1)^k(bc, \omega). \tag{9}$$

The 0-currents (c^k, ω^k) are especially interesting to us since they represent mass distributions. Thus the mass contained in c^k is represented by

$$\int_{c^k} \omega^k. \tag{10}$$

In this way, volume, surface, line and point mass distributions were all contained in a single structure, the 0-currents. The currents thus solved one of the problems which was later solved in the theory of distributions. In his first version of the theory of currents [1936] de Rham was not able to treat more singular distributions than these, which are all measures. In another respect the class of currents was more extensive than the class of distributions since in addition to the 0-currents it included the 1, 2, ...-currents to which no objects correspond in the theory of distributions. For example, the 1-currents represent electrical currents.

It is interesting to note that although the differential lowers the dimension (p) of the current, it still operates like differentiation does in the theory of distributions. Consider, for example, the current $(\mathbb{R}_+, 1)$ that is intuitively a unit electrical current in a cable along $\mathbb{R}_+$ starting at the origin, i.e. corresponding to the Heaviside function. Taking its differential according to definition (9) one obtains

$$d(\mathbb{R}_+, 1) = (\mathbb{R}_+, 0) + (b(\mathbb{R}_+), 1) = (0, 1), \tag{11}$$

since for simplification $(c, 0)$ is set equal to 0. The result $(0, 1)$ represents a unit mass concentrated at the origin. Thus this calculation corresponds closely to the distribution formula

$$\frac{d}{dx} H = \delta. \tag{12}$$

4. Thus de Rham developed a theory of currents which could solve some of the same problems that the theory of distributions could such as treating general mass distributions and differentiating some nondifferentiable functions. However the method of generalization was entirely different from the functional analytical method used in the theory of distributions. De Rham did indicate the functional nature of the currents however [1936, p. 224]:

> A tout $(n - p)$ courant c^{n-p} correspond une fonctionelle linéaire de p courant
>
> $$F(c^p) = I(c^p, c^{n-p}). \tag{13}$$
>
> [de Rham gave a precise definition of the index $I(c^p, c^{n-p})$.] Cela permet, dans le cas généraux, de déterminer le $(n - p)$-courant c^{n-p} par les valeurs de la fonctionelle correspondante sur un certain ensemble de p-courants. C'est ainsi qu'une p-forme ω est déterminée par les valeurs de l'intégrale $\int_c \omega$ pour tout champ.

In [1950] de Rham altered his definition of currents inspired by Schwartz' theory of distributions:

> Dans une variété à n dimensions V, un courant est une fonctionelle $T(\varphi)$, définie sur l'espace vectoriel de toutes les formes φ, C^∞ avec un support compact dans V, qui est linéaire.

Formula (13) offers the possibility of including de Rham's earlier currents in the new structure. Yet the new concept of currents was much more comprehensive than the older one and included distributions which are the 0-currents.[3] More generally de Rham could show that in a suitable sense a current is a differential form with distributions, instead of functions, as coefficients [De Rham 1955, p. 42]. Thus the theory of distributions played a decisive role in the history of de Rham's currents.

Conversely the currents were a factor in the creation of distributions. They did not motivate Schwartz' work, but early in his process of discovery he saw the connection with de Rham's earlier methods, an insight which reinforced his belief in the value of the theory and supported his further work in it.

Chapter 6

Schwartz' Creation of the Theory of Distributions

1. Laurent Schwartz (born 1915) graduated in mathematics from the Ecole Normale Supérieure in 1937. After military service he began his research in 1940 in the Strasbourg science faculty which had fled to Clermont–Ferrand during the German occupation. There he received his doctoral degree in 1943. In 1944 he became an educational assistant at the University of Grenoble, and in 1945 professor in the science faculty at Nancy (see Schwartz' [1949] description of mathematics in France during the war). In 1953 he became attached to the Paris educational system, first (1953–1959) as a professor at the science faculty and later (from 1959) as a professor at the Ecole Polytechnique. Schwartz is considered one of the leading mathematicians in France, maybe of the whole world, and has been a member of the team writing N. Bourbaki's *Eléments de Mathématique.* He has been awarded honorary doctorates at universities all over the world and has received various prizes, most important of all the Field's Medal which he received in 1950 for his creation of the theory of distributions. Laurent Schwartz is known for his political activism. According to himself his strong left wing opinions are guided by his desire to be as rigorous and logical in politics as in mathematics. This explains the influence of his mathematics on his political ideas. The influence of his political convictions on his mathematical work is restricted, he jokingly told me [1978, Interview], to the distraction that his political activity causes on his mathematical research.

At the Ecole Normale Schwartz had three teachers who had played roles in different parts of the prehistory of the theory of distributions, namely, J. Leray, P. Levy and J. Hadamard. From Leray he learned about generalized solutions to partial differential equations (cf. Ch. 2, §46). About Levy he wrote, "P. Levy a eu dès lors une grande influence sur moi, tant pour les probabilités que pour l'analyse classique" [Schwartz 1974, p. 4]. However, Schwartz did not learn about Levy's version of the operational calculus [Levy 1926] (see [Lützen 1979, V, §2]). On the other hand he did learn about Hadamard's partie finie [1978, Interview]. Hadamard and his colleagues,

S. Mandelbrojt and G. Valiron influenced Schwartz' doctoral thesis [Schwartz 1943] done in 1942 at the university in Clermont-Ferrand. "Celle-ci, consacrée a l'étude des sommes d'exponentielles, utilise des méthodes d'analyse fonctionnelle d'une manière nouvelle pour résoudre des problèmes d'approximation de type classique." [Schwartz 1974, p. 5.) This idea of applying functional analysis to classical problems became very important in the theory of distributions. Schwartz was led to this line of investigation by the Bourbaki school which was represented in Clermont-Ferrand by J. Dieudonné, H. Cartan, A. Weil, and others.

> La rencontre de N. Bourbaki m'a initié à des idées toute nouvelles après ma formation d'analyste classique et m'a orienté vers l'algèbre et la topologie, pas tellement pour elles-mêmes que pour leurs applications à l'analyse. Le cours d'analyse fonctionnelle de J. Dieudonné a été à l'origine de ma thèse. [Schwartz 1974, p. 4.]

2. At Grenoble Schwartz was isolated. He explains [Schwartz 1974, p. 5]:

> A la fin de la guerre, travaillant tout seul, je me suis fait une théorie complète de la dualité dans les espaces fonctionnelles généraux, théorie qui m'a paru alors sans application et que j'ai gardée pour moi.

Schwartz generalized the theory of duality of Banach spaces to the theory of duality of Fréchet spaces. He introduced neighbourhoods in the Fréchet spaces themselves and defined the strong dual topology on the dual space. One of the new theorems that he could prove was that a Fréchet space is reflexive[1] if and only if the weakly bounded subsets are weakly precompact. The most important examples of Fréchet spaces known to Schwartz were the space of holomorphic functions and the space $\mathscr{E}$ (or C^∞). Schwartz' results were similar to the results obtained by Mackey (see [Mackey 1943, 1946] who was unknown to Schwartz. However, Schwartz' work did not form a comprehensive theory as did Mackey's, but rather had the character of separate results (all this information is from [1978, Interview]). This work of Schwartz' on abstract functional analysis was never published, but it had a remarkable effect on the creation of the theory of distributions:

> elle [this previous abstract work] devait être la clef de la théorie des distributions. C'est cette formation antérieure qui fait que la "découverte" des distributions fut en fait presque instantanés au début de 1945. [Schwartz 1974, p. 5.]

Schwartz' "discovery" of the theory of distributions took approximately half a year, a period which is certainly instantaneous compared with the period of time in which concrete problems were in severe need of the theory. On the other hand, it is such a decisive instant in the solution or simplification of these problems that it is worth while to look more closely at this moment. It will then be seen that Schwartz' creation was accomplished in two steps and that the theory of duality played a direct role only in the last step.

3. The starting point for Schwartz' creation of the theory of distributions was his work on generalized solutions of partial differential equations. This work in turn did not directly stem from any of the problems considered in Ch. 2, but was brought about by Choquet and Deny's paper "Sur quelques propriétés de moyenne caractéristiques des fonctions harmoniques et polyharmoniques" [1944]. One of the main theorems of this paper was the following:

(A) *Let F be a function of n variables, continuous in a domain D. If there exists a mass distribution, λ_0, supported by a compact subset E_0 of $\mathbb{R}^n$ such that*

$$\int_E F\,d\lambda = 0$$

for all mass distributions (E, λ) similar[2] *to (E_0, λ_0), then F is polyharmonic in D.*[3]

In the last part of their article Choquet and Deny suggested an application of this theorem

> que M. H. Cartan a eu l'obligeance de nous signaler, relativement à l'approximation uniforme de toute fonction continue définie sur un compact fixe, par des combinaisons linéaires de fonctions déduites très simplement d'une fonction continue donnée.

In order to state this application precisely one must define the abovementioned set of functions called $\mathscr{F}$ which are "déduites très simplement d'une fonction continue donnée" called F: For every similitude $S(\mathbb{R}^2 \to \mathbb{R}^2)$ define F' such that $F'(S(M)) = F(M)$ for all $M \in \mathbb{R}^2$. $\mathscr{F}$ consists of all such functions F'. Cartan's theorem then states:

(B) *Si la fonction continue F n'est pas polyharmonique dans D, toute fonction Φ définie et continue sur E_0 peut être approchée uniformement par des combinaisions linéaries finies de fonctions de $\mathscr{F}$.*

This theorem follows directly from Choquet and Deny's theorem (A) by using the Hahn–Banach theorem and Riesz representation theorem. Indeed from theorem (A) it follows that if $F(M)$ is continuous and not polyharmonic, then there does not exist a mass distribution (E_0, λ_0) for which

$$\int_{E_0} F'\,d\lambda_0 = 0$$

for all $F' \in \mathscr{F}$. According to Riesz' representation theorem, this amounts to saying that there does not exist a nonzero continuous linear functional U on $C(E_0)$ for which

$$U(F') = 0 \qquad \forall F' \in \mathscr{F}.$$

The Hahn-Banach theorem then shows that the span of $\mathscr{F}$ is dense in $C(F_0)$ (we say that $\mathscr{F}$ is fundamental).

4. Immediately following Choquet and Deny's article in the *Bull. Soc. Math. France* was a paper, "Sur certaines familles non fondamentales de fonctions continues", by Schwartz [1944] in which he generalized Cartan's theorem (B). Deny and Choquet had made minor generalizations, among which they remarked that in the formation of $\mathscr{F}$ it was enough to perform infinitely many rotations around a point M (as opposed to all rotations). Schwartz left out all of the rotations and retained only the translations and the multiplications in the definition of $\mathscr{F}$. This invalidated the conclusion of the theorem (B) as stated by Cartan, but he could prove the following:

(C) *Pour que le système de fonctions continues*

$$U(\lambda x_1 + \xi_1, \lambda x_2 + \xi_2, \ldots, \lambda x_n + \xi_n)$$

des variables $x_1, \ldots, x_n$ *soit non fondamental sur un compact K de l'espace à n dimensions, lorsque* λ *et* $\xi_1, \xi_2, \ldots, \xi_n$ *prennent toutes les valeurs réelles possibles, il faut (et il suffit si l'intérieur de K n'est pas vide) que U soit solution généralisée d'au moins une équation aux dérivées partielles du type*

$$0 = \sum_{p_1 + p_2 + \cdots + p_n = m} a_{p_1 p_2 \cdots p_n} \frac{\partial^m U}{(\partial x_1)^{p_1} (\partial x_2)^{p_2} \cdots (\partial x_n)^{p_n}}, \tag{1}$$

where $a_{p_1} \ldots {}_{p_n}$ *are constants.*

Schwartz defined $U(x_1, \ldots, x_n)$ as a generalized solution of equation (1) if it was a uniform limit of ordinary solutions on every compact subset of $\mathbb{R}^n$. Thus Schwartz like Sobolev started with a sequence definition of a generalized solution. However, the type of convergence he used differed from the L^1 convergence used by Sobolev. Schwartz remarked that the wave equation

$$\frac{\partial^2 U}{\partial x^2} - \frac{\partial^2 U}{\partial y^2} = 0$$

has the generalized solutions $f(x + y) + g(x - y)$ for all continuous functions f, g. He also remarked that all generalized solutions to Laplace's equation are ordinary solutions.

5. In order to see how Schwartz was inspired to take the next step toward distributions, I shall consider the proof of the necessity assertion of the above theorem (C).

If the system is not fundamental then according to the Hahn–Banach theorem and Riesz' representation theorem there exists a mass distribution $\Phi(x_1, \ldots, x_n)$ different from zero such that

$$\int U(\lambda x_1 + \xi_1, \ldots, \lambda x_n + \xi_n)\, d\Phi(x_1, \ldots, x_n) = 0$$

for all $\lambda, \xi_1, \ldots, \xi_n \in \mathbb{R}$.

Functions V (which will play the role of the approximating sequence) are constructed from the convolution integral (convolution with $\rho(-u)$)

$$\begin{aligned} V(x_1, \ldots, x_n) &= \int U(x_1 + u_1, \ldots, x_n + u_n) \rho(u_1, \ldots, u_n)\, du_1, \ldots, du_n \\ &= \int U(u_1, u_2, \ldots, u_n) \rho(u_1 - x_1, \ldots, u_n - x_n)\, du_1, \ldots, du_n. \end{aligned}$$

If ρ is taken to be a C_c^∞ function, the integral is defined and V is differentiable infinitely often. Moreover, V satisfies the equation

$$\int V(\lambda x_1 + \xi_1, \ldots, \lambda x_n + \xi_n)\, d\Phi(x_1, \ldots, x_n)$$

$$= \int [\rho(u_1, u_2, \ldots, u_n)\, du_1, \ldots, du_n$$

$$\times \int U(\lambda x_1 + \xi_1 + u, \ldots, \lambda x_n + \xi_n + u_n)\, d\Phi(x_1, \ldots, x_n)] = 0. \quad (2)$$

Since the system of continuous functions, $x_1^{p_1}, x_2^{p_2}, \ldots, x_n^{p_n} (p_i \in \mathbb{N})$ is fundamental on a compact set, there exists a system of natural numbers $q_q, q_2, \ldots, q_n$ such that

$$\int x_1^{q_1} x_2^{q_2}, \ldots, x_n^{q_n}\, d\Phi(x_1, x_2, \ldots, x_n) \neq 0. \quad (3)$$

Differentiating the left-hand side of (2) m times (where $\sum q_i = m$) with respect to λ Schwartz got for $\lambda = 0$:

$$0 = \sum_{p_1 + p_2 + \cdots + p_n = m} \left[\frac{m!}{p_1! p_2! \cdots p_n!} \frac{\partial V(\xi_1, \ldots, \xi_n)}{(\partial \xi)^{p_1} \cdots (\partial \xi_n)^{p_n}} \right.$$

$$\left. \times \int x_1^{p_1} \cdots x_n^{p_n}\, d\Phi(x_1, \ldots, x_n) \right],$$

so that for

$$a_{p_1, p_2, \cdots, p_n} = \frac{m!}{p_1! p_2! \cdots p_n!} \int x_1^{p_1} x_2^{p_2} \cdots x_n^{p_n}\, d\Phi(x_1, x_2, \ldots, x_n), \quad (4)$$

(which are not all zero according to (3)) V is a solution to the equation (1). By choosing a family of functions $\rho(u_1, u_2, \ldots, u_n)$ which are supported by neighbourhoods tending to zero and which all satisfy

$$\int \rho(u_1, u_2, \ldots, u_n)\, du_1\, du_2 \cdots du_n = 1,$$

Schwartz found that the corresponding Vs tend to U uniformly on any compact set. Thus U is a generalized solution of (1) with coefficients (4).

6. A few days after writing the above article Schwartz set himself the task of providing a better definition of a generalized solution.[4] He focused on the fact that the generalized solution U in the proof of the theorem worked as a convolution operator taking a C_c^∞ function (ρ) into a C^∞ function (V).

Thus he introduced the new object which he called a "convolution operator". He defined it as a continuous linear operator T from $\mathcal{D}$ to $\mathcal{E}$ with the property that

$$T \cdot (\varphi * \psi) = (T \cdot \varphi) * \psi \qquad \forall \varphi, \psi \in \mathcal{D}.$$

Here $\mathscr{D}$ is C_c^∞ equipped with the now well-known notion of convergence, i.e. $f_n \to f$ if all f_ns have their support in a fixed compact set K and f_n with all their derivatives converge to f and its derivatives uniformly on K. On $\mathscr{E}$ (C^∞) he used the usual Fréchet topology. The continuity was taken in the sequence-sense, i.e. if a sequence $f_n \to f$ in $\mathscr{D}$ then $T \cdot f_n \to T \cdot f$ in $\mathscr{E}$.

Schwartz immediately saw that the convolution operators gave a generalization of the continuous functions in $\mathbb{R}^n$ since such a function g could be identified with the operator

$$f \rightarrow f * g.$$

He also saw that the δ-function could be given a rigorous definition as the convolution operator

$$\delta \cdot \varphi = \varphi.$$

Since the starting point for the convolution operators was the theory of generalized solutions to partial differential equations, Schwartz was naturally led to the question of defining the derivative of an operator. This he did by the formula

$$\left(\frac{\partial}{\partial x_i} T\right) \cdot \varphi = T \cdot \frac{\partial}{\partial x_i} \varphi,$$

generalizing the formula

$$\frac{\partial}{\partial x_i} \psi * \varphi = \psi * \frac{\partial}{\partial x_i} \varphi.$$

Conversely he defined the primitive of an operator in the obvious way.

He adopted other classical notions, for instance, convolution and multiplication by a function. Convolution of two operators S and T was defined in the obvious way

$$(S * T) \cdot \varphi = S \cdot (T \cdot \varphi).$$

However, multiplication with a C_c^∞ function caused difficulties. Schwartz overcame the problem by first proving that every convolution operator was locally an nth derivative of a continuous function. In this proof he used what he knew about neighbourhoods in $\mathscr{D}_K(\mathbb{R}^n)$ (cf. [Schwarz 1950/51, Vol. I, p. 82, Theorems XX and XXI]) which had come from his earlier work with Fréchet spaces (§2). He then used Leibniz' rule successively to define the product $f \cdot T$ where f is a C_c^∞ function and T a convolution operator.

(1) If T is the derivative g' of a continuous function on the support K of f, he defined:

$$f \cdot T = f \cdot g' \equiv (fg)' - f'g,$$

the right-hand side of which is well defined.

(2) If $T = g''$ he defined

$$f \cdot T = f \cdot g'' \equiv (fg)'' - 2f' \cdot g' - f''g,$$

the right-hand side of which has meaning according to (1).

Continuing in this way, Schwartz could define $f \cdot T$ for all $T = g^{(n)}$ on K, and therefore, by the above finiteness theorem, for all convolution operators. This however gave a very "heavily working definition" and it became a problem (but one he solved) to show that the product was independent of the particular representation of T chosen.

Further he gave a definition of convergence of convolution operators.

7. Schwartz devised all these definitions and several theorems about the convolution operators during one night in October 1944. In the following period of approximately six months he continued to work with the new concept and proved further theorems in the field. Around February 1945, he began to develop a theory of Fourier transformations. However, there he met with great difficulties with the unwieldy operators. He struggled with these difficulties for some months. Then one day, probably in April 1945, in his office in Grenoble, where he had become Chargé d'enseignement, he suddenly realized that these problems could be overcome if he defined his generalized functions not as operators but as functionals which he called distributions. When he first arrived at this idea, he considered it so obvious that later he thought it was "stupid" that he had not seen the superiority of this definition immediately in October 1944. "It was visible but I did not see it" he told me in the [1978 Interview]. He thought that the difficulty behind such innovations is the "cancellation of the resistance" to innovations.

According to his own statement, there were two facts which could have suggested the "real" definition of distributions:

(1) His earlier work on the duality of Fréchet spaces provided him with an abstract theory of continuous linear functionals on $\mathscr{E}$, without a "concrete" representation with which he could calculate. From October 1944 he also had the convolution operators with compact support: $\mathscr{E} \rightharpoonup \mathscr{E}$. However, he did not combine the two ideas before the spring of 1945.

(2) He knew that measures, especially the δ-measure, could be represented as functionals.

During the winter of 1944-45 Schwartz continually told H. Cartan about his progress with the convolution operators. After he had changed his objects to functionals instead of operators, he immediately informed Cartan of the innovation. Cartan apparently found the idea very obvious, for according to Schwartz [1978, Interview] he responded with a very French "Ah!" meaning: "Of course. Why had I not thought about that before?"

8. From the spring of 1945 Schwartz developed his new theory of distributions. In the theory of operators his earlier abstract functional analytic work had already played a role, for instance, for the proof of the finiteness theorem. Now in the theory of distributions it occupied a much more central position and aided its development considerably. However, the influence did not proceed in one direction only, from abstract theory to the theory of distributions; it went the other way as well: for example, the concept of an inductive limit of Fréchet spaces had its roots in the theory of distributions. Schwartz knew very well that convergence in $\mathscr{D}$ could not be obtained from a Fréchet topology on that space. In his first work on distribution theory he, therefore, did not introduce any system of neighbourhoods in $\mathscr{D}$ but contended himself with introducing bounded sets in $\mathscr{D}$ which in turn allowed him to define the dual topology on $\mathscr{D}'$. When J. Dieudonné heard Schwartz' topological description of the space $\mathscr{D}$, he got his idea for the abstract theory of inductive limit spaces. In 1949 Dieudonné and Schwartz [1949] wrote a joint article on the subject in which they proved the main theorems which Schwartz had used and proved for the concrete case of the spaces $\mathscr{D}$ and $\mathscr{D}'$.

> C'est à partir de notre travail que N. Bourbaki a introduit les notions d'espace bornologique, espace tonnelé, etc.... [Schwartz 1974, p. 20.]

Thanks to his previous work on functional analysis, Schwartz developed his new theory of distributions so rapidly that he was able to lecture on it in the winter of 1945/46 at the *Cours Peccot* at the *Collège de France* in Paris. In response to these lectures there were many comments on applications and precursors of his theory. Before entering into more detail on this, I want to list the parts of the "prehistory" which had inspired Schwartz and the parts of which he was ignorant.

9. From the above description of Schwartz' route to the theory of distributions, it is clear that the theory of generalized solutions to partial differential equations was his main source of inspiration. He wrote [Schwartz 1974, p. 5]:

> Je savais déja [in 1944] que les solutions d'une équation aux dérivées partielles elliptique pouvaient se définir sans mettre de dérivées dans la définition et qu'elles étaient automatiquement indéfiniment dérivables, et qu'au contraire, pour les équations hyperboliques, it devait y avoir une définition n'utilisant pas de dérivées et donnant des solution effectivement non dérivables.

He had learned parts of this theory at the lectures which J. Leray (Ch. 2, §46) had presented at the Collège de France. In particular, he was aware of Riemann's generalized second-order differentiation in his work on trigonometric series (see Ch. 2, §16).

Three other facts played a part in leading Schwartz to distributions: functional analysis, the δ-function, and de Rham's currents. Functional analysis was the basis for the entire formalism, both for the convolution operators and more directly for the distributions. The δ-function and the currents were problems which Schwartz had in mind when he developed the theory of operators and which he immediately saw how to fit into the new theory. He had noted the mathematical insufficiency of the δ-function during his student days. However, he did nothing to solve the problem then, but it continued to disturb him.

He had learned about de Rham's currents in 1942 from de Rham personally, who had vaguely indicated the functional character of his currents and their connection to measures. These ideas had impressed Schwartz so much that he immediately saw that his distributions could be used in the theory of currents. However, he had only an intuitive idea about the connection between his theory and de Rham's, and he postponed more detailed investigations. Before he had the chance to accomplish anything more concrete, de Rham, who had learned about the theory of distributions, had already redefined his currents in terms of distributions in 1950 (Ch. 5, §4).

Although Schwartz was aware of Hadamard's partie finie, it played no role in the creation of the theory of convolution operators or distributions.

Schwartz also knew about some nonrigorous works on generalized Fourier transforms: tables giving generalized functions as Fourier transforms (see Ch. 4, §27).

10. In 1944 Schwartz was unfamiliar with the rest of what I have called the prehistory of the theory of distributions. Thus he was unaware of:

(1) Bochner's and Carleman's theories of generalized Fourier transformations.
(2) Sobolev's work.
(3) The Heaviside calculus.
(4) The physicists' use of improper functions more complex than the δ-function.
(5) Fantappié's analytical functionals.

He later learned about these previous formations from colleagues. However, by then he had already developed his theory so far that there was nothing new to be learned from them, at least nothing which could inspire him to new techniques or concepts in the theory of distributions. In 1946 Leray called Schwartz' attention to Sobolev's work. From his audience at the Peccot course he received other leads to the work of his predecessors. The electrical engineers in particular confirmed his belief in the importance of the theory of distributions. They attended the course not because *Schwartz* had pointed out in his announcement the importance of the theory to electrical engineering, but because *they* could infer from the description of the contents that the course would be of interest to them. They told Schwartz about the operational calculus and encouraged him to continue his work with the theory especially with the treatment of the Fourier and Laplace transformations and the convolution. Moreover, they asked him to make his planned book on the subject so elementary that they would be able to understand it.

The interest shown by the electrical engineers caused Schwartz to think that the main application of the theory of distributions would be in the field of electrical engineering. He knew, of course, from the beginning that distributions were important in the solution of partial differential equations. Only later, however, did he see that their importance was so far-reaching.

The significance which Schwartz attached to the application of the theory of distributions in operational calculus can be seen in his decision to give a talk at a conference of the Societé des Radioélectriciens in December 1946. An account of this was published in [1948] in the *Annales des Télecommunications.*

11. This article was the last of four articles on the theory of distributions which Schwartz wrote before his monograph *Théorie des Distributions* was published in 1950/51. The first article [1945] and this last one [1948] both gave general surveys of the theory and contained its main ideas: distributions, multiplication, Fourier and Laplace transformation, integration of distributions and solution of differential equations in $\mathscr{D}'$. The first article places more emphasis on the purely mathematical structure of distributions while the last is more concerned with applications. The second and third papers [Schwartz 1947b, 1947/48] dealt primarily with the Fourier transformation. In both the first and the fourth of Schwartz' papers [1945, 1948] the Fourier transform was only defined for slowly increasing functions by the definition

$$\mathscr{F}(f)(x) = \lim_{A, B \to \infty} \int_{-A}^{B} f(\omega) \exp(2\pi i \omega x)\, d\omega,$$

where the limit is taken in $\mathscr{D}'$. The idea of the generalized Fourier transform defined as the dual of the ordinary Fourier transform is not present in these two articles.

However in [1945] Schwartz remarked:

> Mais il est également possible d'élargir beaucoup le champ d'application des transformations de Fourier et Laplace, et de définir les transformées de toutes les distributions, quels que soient leur irrégularité et leur comportement à l'infini; on est alors obligé d'introduire une nouvelle famille de distributions d'un maniement nettement plus compliqué et moins intuitif.

This strongly indicates that Schwartz did not possess the idea of the Fourier transformation of tempered distributions in 1946.[5] However, later [1978, Interview] he thought—although he was not quite sure—that he was close to this idea already during the time when he was trying to define Fourier transformations for convolution operators. In any case, it is certain that in June 1947, he had the tempered distributions and the corresponding Fourier transformation, because during the week of June 15–22 he spoke on the "Théorie des distributions et transformation de Fourier" [Schwartz, 1947b] at a congress on harmonic analysis at Nancy, and developed the basic elements of this theory. He described the spaces $\mathscr{S}$ and $\mathscr{S}'$, and the

Fourier transformation in these, in a way which has now become standard. He called the distributions in $\mathscr{S}'$ "spherical distributions" because:

> Pour qu'une distribution T de $(\mathscr{D}')$ distribution sur $\mathbb{R}^n$ appartienne à $(\mathscr{S}')$, il faut et il suffit q'elle soit prolongeable en une distribution sur la sphère S^n,

where $\mathbb{R}^n$ has been identified with $S^n\setminus\infty$.

In his monograph *Théorie des Distributions* he did not attach much importance to the above characterization of distributions in $\mathscr{S}'$, but stressed that $\mathscr{S}'$ consisted of the derivatives of functions of slow growth at infinity (i.e. continuous functions growing more slowly at infinity than some polynomial). Therefore he changed the name of the members of $\mathscr{S}'$ to temperate distributions but retained the name $\mathscr{S}'$ for the space.

12. Schwartz' monograph, which was published in two volumes in 1950/51, immediately became the standard work in the theory of distributions. He has continuously brought the books up to date and has thereby succeeded in maintaining their completeness so well that Treves who among others wrote on the subject could characterize *Théorie des Distributions* as "still the best and most comprehensive exposure of the theory" [Treves 1975, p. 465].

Schwartz continued to work on the theory of distributions after publication of the *Théorie des Distributions.* The first great success of the theory not contained in the textbook was the rigorous formulation and proof of the kernel theorem. It was Dirac who inspired Schwartz to the proof of this theorem (see Ch. 4, §29, especially the quotation).

In connection with the kernel theorem Schwartz extended distributions to vector-valued distributions. He applied distributions to a theory of elementary particles [1969] and extended the theory of Radon measures. For a more detailed account of Schwartz' vast mathematical work after 1950 the reader should consult Schwartz' autobiography [1974].

Concluding Remarks

The Concluding Remarks have three parts. In §1 a question from the Preface is discussed §2–4 treat the reception of the theory of distributions and the different views upon the theory. Finally, in §5–7 an attempt is made to place the prehistory of the theory of distributions in the general history of mathematics.

1. In the Preface I posed the questions: Who invented distributions and when? and I gave the provisional answer: Sobolev in 1936 and Schwartz in 1950. After having discussed the prehistory of the theory of distributions in detail the question seems too general and needs specification. If one asks about the first people to use distributions in mathematics, the answer is Fourier 1822, Kirchhoff 1882 and Heaviside 1898. If one asks for a rigorous theory, which possibly only implicitly used distributions, the answer is Bochner 1932. If one wants to know who first defined distributions rigorously as functionals, the answer is Sobolev 1935 and finally, if one wants to point to the person who saw the far-reaching applications of distributions and created a broad theory of these objects, Schwartz is the one to cite with 1945–1950 as his years of publication. In Soviet and Eastern European texts on the theory of distributions the third of these questions is usually stressed so that Sobolev becomes the hero; in Western texts the credit is often given to Schwartz because only the last question is asked (see, for example, Dieudonné [1964]). The many answers to the question from the Preface reflects the ambiguity in the term "discoverer". Schwartz himself has [1978, Interview] drawn my attention to the book, *Democracy Ancient and Modern* [Finley 1973, pp. 13–14] in which M. Finley discusses the nature of discovery.

> It was the Greeks, after all, who discovered not only democracy but also politics, the art of reaching decisions by public discussion and then of obeying those decisions as a necessary condition of civilized social existence. I am not concerned to deny the possibility that there were prior examples of democracy, so-called tribal democracies, for instance, or the democracies in early Mesopotamia that some

> Assyriologists believe they can trace. Whatever the facts may be about the latter, their impact on history, on later societies, was null. The Greeks, and only the Greeks, discovered democracy in that sense, precisely as Christopher Columbus, not some Viking seaman, discovered America.
>
> The Greeks were then—and this no one will dispute—the first to think systematically about politics, to observe, describe, comment and eventually to formulate political theories.

According to this theory of discovery, Schwartz is evidently the discoverer of the theory of distributions, since he was the first to see the full consequences of his theory and to make a strong influence on the later development of the theory. By choosing 1950 as the boundary between the prehistory and the history of the theory of distributions, I have used a definition of discovery similar to Finley's. However, just as the Vikings' discovery of America was a great event, so Sobolev's definition and use of distributions is a highlight of the prehistory of distributions.

This discussion reveals, as is often the case with such simple questions in the history of mathematics, that asking who and when in connection with the history of distributions does not make sense in its broad generality. More specifically, however, one can state that Sobolev invented distributions, in the modern sense, and Schwartz created the *theory* of distributions.

2. Laurent Schwartz' theory of distributions was well received both by mathematicians and by physicists who could then use improper functions in good conscience. In November 1947, H. Bohr wrote from Copenhagen to his former student A. Aaboe, now professor at Yale:

> ... we have had several extremely interesting visits by foreign mathematicians, in the first line the young French mathematician Prof. Laurent Schwartz; I intend to propagandize strongly for his eminent contribution to the classical differential and integral calculus in the United States, which however may prove unnecessary.[1]

Partly because of H. Bohr's "propaganda", L. Schwartz was in 1950 awarded the Fields' Medal, the highest honour one can receive in mathematics. In a speech given on behalf of the committee to select the Fields medalists H. Bohr said:

> ... one of the greatest merits of Schwartz' work consists ... in his creation of new and most fruitful notions adapted to the general problems, the study of which he has undertaken. While these problems themselves are of classical nature, in fact dealing with the very foundation of the old calculus, his way of looking at the problem is intimately connected with the typical modern development of our science with its highly general and often very abstract character. Thus once more we see in Schwartz's work a confirmation of the words of Felix Klein that great progress in our science is often obtained when new methods are applied to old problems. [Bohr 1950.]

After reviewing the central ideas in the theory of distributions, H. Bohr concluded:

> Schwartz is now preparing a larger general treatise on the theory of distributions, the first, very rich, volume of which has already appeared. In his introduction to this treatise he emphasizes the fact that ideas similar to those underlying his theory have

earlier been applied by different mathematicians to various subjects—here only to mention the methods introduced by Bochner in his studies on Fourier integrals—and that the theory of distributions is far from being a "nouveauté révolutionnaire". Modestly he characterizes his theory as "une synthèse et une simplification". However as in the case of earlier advances of a general kind—to take only one of the great historic examples, that of Descartes' development of the analytic geometry which, as is well-known, was preceded by several analytic treatments by other mathematicians of special geometric problems—the main merit is justly due to the man who has clearly seen, and been able to shape, the new ideas in their purity and generality.

No wonder that the work of Schwartz has met with great interest in mathematical circles throughout the world, and that a number of younger mathematicians have taken up investigations in the wide field he has opened for new researches. [Bohr 1950.]

In [1948] A. Weil had already emphasized the importance of the theory of distributions:

Il y a là [in the theory of distributions] peut-être le principe d'un calcul nouveau, reposant en définitive sur le théorème de Stokes généralisé, et qui nous rendra accessible les relations entre opérateurs différentiels et opérateurs intégraux.... Dans ces recherches, on voit peut-être s'ébaucher un calcul opérationnel destiné à devenir d'ici un siècle ou deux un instrument aussi puissant que l'a été pour nos prédécesseurs et pour nous-mêmes le calcul différentiel.

J. Dieudonné many times stressed the great success of the theory of distributions. In [1964], for example, he wrote:

The applications of these new ideas [the theory of distributions], and in particular the extended range offered to the convolution product and the Fourier transformations, were not long in making themselves felt; I need only mention here the work of Gårding, Hörmander, Malgrange, Ehrenpreis, Lojasiewicz, Calderon and many others, which has taught us so much on the general properties of linear partial differential equations, especially on existence and uniqueness problems, now essentially solved for systems of arbitrarily high order with constant coefficients.

Thus the theory of distributions gained acceptance very rapidly as an extremely significant mathematical innovation.

3. However skepticism concerning the use of generalized functions, in particular distributions, was voiced by R. Courant. In the English edition of *Methods of Mathematical Physics, II* [Courant–Hilbert 1962] Courant tempered his generally positive attitude towards the theory of distributions with the following warnings [p. 768, note 3]:

Formal simplifications thus attainable must not create the illusion that the core of intrinsic difficulties can thereby be mastered rather than merely isolated and clarified. Often the genuine difficulty is shifted to the final task of ascertaining in what sense a result obtained in terms of ideal functions is indeed expressible by ordinary functions.

He added [Courant – Hilbert 1962, p. 788]:

Introducing ideal functions may appear a sweeping extension of ordinary calculus. Yet, in the realm of ideal functions not all operations of classical calculus can be carried out. Thus the advantage of securing unrestricted differentiability is partly offset by the loss of freedom in multiplying functions or in forming composite functions. It is not even completely true that an ideal function of several variables becomes an ideal function of fewer variables if some of the others are kept constant in a domain of definition.

More recently and in a more general form the importance of the theory of distributions to the *development* of modern mathematics has been questioned by F. E. Browder. In a talk, "The relation of functional analysis to concrete analysis in 20th century mathematics" [1975], he characterized the theory of distributions as an important and curious turning point and said that "it provided a relatively useful general language for communication between analysts and applied mathematicians". "However", he continued, stressing the negative aspects of the theory,

> one cannot say that the theory of distributions plays the same role as spectral theory, because the theorems in the theory of distributions do not seem to have any specific power of their own, although, the framing of problems in terms of distribution theory seems to have had a very important organizing role. . . .
>
> The theory of distributions has provided a language rather than a methodology. It is used as a way to organize and to state problems, in a more general and flexible form; then, other tools are applied.

4. Browder's point of view—that the theory of distributions is primarily a language—is not so negative a charge considering the importance it has had in analysis and applied mathematics.

However it has been argued that the theory of distributions may not be an appropriate language. Many alternative definitions or even theories of generalized functions have been given, the inventors of which claim their superiority over the theory of distributions (see Appendix).

Courant [Courant–Hilbert 1962] felt that the best generalization of the concept of function may not yet have been found. Thus, after introducing distributions in three equivalent ways [pp. 775–781] he remarked [p. 798]:

> Notwithstanding the merits of the theory developed in this appendix, the above remark should call attention to the need for further study of other less well explored possibilities of generalizing the concept of function by introducing suitable ideal elements. The value of such concepts should be measured not by their formal generality but by their usefulness in the broader context of analysis and mathematical physics.

However, I have the impression that Dieudonné spoke for the majority of mathematicians when he said in [1964] concerning Schwartz' approach to the theory of distributions:

> although many other approaches to distribution theory have since been proposed, none offers, in my opinion, the flexibility and power of the original description of Schwartz.

Today Schwartz' theory of distributions is by far the most applied theory of generalized functions. Only the theory of hyperfunctions and perhaps the theory of nonstandard functions are likely to threaten its prominent position in the near future.

5. The prehistory and the creation of the theory of distributions illustrate many general patterns characteristic of the history of mathematics. The underlying ideas for the theory of distributions had, as is the case with most mathematical innovations, already "been in the air" for some time. For this reason it is not surprising that distributions were invented independently by Sobolev and Schwartz (and in a different form by Tolhoek and perhaps Bochner and Carleman). This fact only corroborates M. J. Crowe's eighth "law" concerning conceptual change in mathematics [Crowe 1975]:

> Multiple independent discoveries of mathematical concepts are the rule, not the exception.

The way in which these ideas "in the air" grew into the mature theory of distributions also follows well-known patterns. The most conspicuous of these is Schwartz' fusion of several tricks, methods, notions and ideas. Dieudonné has described this as the typical pattern of innovation in mathematics [1975, p. 537]:

> progress in mathematics results, most of the time, through the imaginative fusion of two or more apparently different topics.

In an earlier paper [1964] Dieudonné compared this aspect of the prehistory of the theory of distributions with the creation of the calculus by Newton and Leibniz (see quotation in the Introduction, §3).

Thus the development of the theory of distributions fits nicely into the category of "Fusion" in E. Koppelman's [1975] taxonomy of progress in mathematics. However, it belongs equally well in the category of "Transplantation", since distribution theory was created by transplanting functional analytical ideas into concrete analysis. Koppelman states, however, that "transplantation always leads to growth of the borrowing field, but has little immediate effect on the field from which the technique was borrowed". This particular characteristic of "transplantation" does not apply to the development of the theory of distributions since growth in both the borrowing field (concrete analysis) and the field from which the ideas were borrowed (functional analysis) was stimulated: the invention of LF spaces was an immediate consequence of the creation of distribution theory and in a less immediate way the whole theory of topological vector spaces was strongly motivated by distributions.

The "transplantation" aspect of the creation of distributions was stressed by H. Bohr in the quotation above (§2), where he cites F. Klein's remark "that great progress in our science is often obtained when new methods are applied to old problems". The discoveries of both Sobolev and Schwartz were transplantations of functional analysis to concrete analysis, but only Schwartz' theory of distributions was a fusion of problems in concrete analysis.

6. It is clear from the above discussion that the prehistory of the theory of distributions confirms many of the patterns associated with developments in the history of mathematics. How, then, does the last 50 years of the prehistory more specifically fit into the mathematical development in the first half of this century? It is difficult to answer this question since the history of modern mathematics has not been studied adequately enough to produce a clear picture of its characteristic features. However our present knowledge allows some general remarks.

It has often been emphasized that mathematics in this century is characterized by diversity and overspecialization. H. Weyl has expressed it in the following way:

> Whereas physics in its development since the turn of the century resembles a mighty stream rushing on in one direction, mathematics is more like the Nile delta, its waters fanning out in all directions. [Weyl 1951, p. 523.]

However, Dieudonné has recently expressed the view that "mathematics is more unified than it has ever been before" [1975, p. 537], a point of view he shares with many colleagues. Unification of diverse fields have been accomplished by fusing seemingly unrelated theories, usually on a higher level of abstraction.[2] The theory of distributions is no exception.

In this century the unification of different mathematical theories is usually done by creating an axiomatic system, the axioms of which are satisfied by these theories. Important examples are topology and different algebraic structures. In the axiomatic treatment of mathematics one focuses on the operations and on the laws satisfied by these operations, rather than on the specific nature of the mathematical objects involved. This "structural" quality is often emphasized as the property most characteristic of twentieth-century mathematics. The theory of distributions is *not* structural mathematics in this sense. To be sure the theory of distributions is built on the highly axiomatized mathematical field of functional analysis, but it is not axiomatic in its construction. It unites different mathematical methods and theories not by imbedding them in an axiomatic structure in which the nature of the elements is irrelevant, but precisely by constructing new mathematical objects: the distributions. The operations for these mathematical objects have not been unified in the theory of distributions. For example, the differential operators and the Fourier transformation, which were the basic operations in two of the different theories united in the theory of distributions, continue to be distinct operations which operate in the *same* domain: the space of distributions.

The prehistory of the theory of distributions reflects the nonstructural character of the subject. The few connections made between its different trends before 1945 were all nonstructural. For example, the comparisons between the different types of generalized solutions or the more implicit links drawn on the basis of the Stieltjes integral did not bear the mark of structuralism. Neither did Schwartz' final fusion of the different trends. Hence the history told in this book is not representative of the distinctive structural mark of the development of modern mathematics.[3]

7. Another characteristic which distinguishes the latter part of the *prehistory* of the theory of distributions from the development of mathematics in the first half of the twentieth century is its close relationship to physics. The purity of mathematics in this century has been stressed by many mathematicians. After outlining what he found to be the most essential developments in recent mathematics, Dieudonné [1964] concluded:

> As a final remark, I would like to stress how little recent history has been willing to conform to the pious platitudes of the prophets doom, who regularly warn us of the dire consequences mathematics is bound to incur by cutting itself off from the applications to other sciences. I do not intend to say that close contact with other fields, such as theoretical physics, is not beneficial to all parties concerned; but it is perfectly clear that of all the striking progress I have been talking about, not a single one, with the possible exception of distribution theory, had anything to do with physical applications; and even in the theory of partial differential equations, the emphasis is now much more on "internal" and structural problems than on questions having a direct physical significance.[4] Even if mathematics were to be forcibly separated from all other channels of human endeavour, there would remain food for centuries of thought in the big problems we still have to solve within our own science.

There is no doubt whatsoever that the development of the theory of distributions occurred in much more *direct* contact with physics than most other innovations in the twentieth century. On the other hand, in my opinion, modern mathematics owes much more to physics through an *indirect* contact than Dieudonné suggests.[5] However, it is not possible to verify this on the basis of the existing studies in the history of recent mathematics. Nor is it possible at present to carry the comparison between the general history of twentieth-century mathematics and the prehistory of distributions in this period much further than it has been done here. It is to be hoped that a general history of recent mathematics will soon be possible with the help of more "case studies" in this field.

Appendix

Alternative Definitions of Generalized Functions

Distributions can be defined in three essentially different ways: as functionals, sequences or improper derivatives. These three different approaches are described in §1–3. The next three sections describe three more generalization methods, which yield objects that are not equivalent to distributions: Mikusiński's operators, hyperfunctions and nonstandard functions. A more detailed account of the different approaches can be found in [Temple 1953, 1955], [Naas and Schmid 1961], [Slowikowski 1955] and [Beltrami 1963] and in the references below.

1. Functionals. A sequence $\varphi_\nu \in C_c^\infty(\mathbb{R}^n)$ is said to converge to 0 if all the φ_νs have their supports contained in *one* compact set K (independent of n) and $\varphi_\nu \to 0$ uniformly together with all its derivatives (this convergence can be defined by an LF topology). The space C_c^∞ with this notion of convergence (topology) is called $\mathscr{D}$. A continuous linear functional on $\mathscr{D}$ is called a distribution. A locally integrable function f is identified with the distribution T defined as

$$T(\varphi) = \int_{\mathbb{R}^n} f \cdot \varphi. \tag{1}$$

The derivative $(\partial/\partial x_i)T$ of a distribution T is defined as

$$\frac{\partial}{\partial x_i} T(\varphi) = -T\left(\frac{\partial}{\partial x_i} \varphi\right). \tag{2}$$

A sequence of distributions T_i is said to converge to T in $\mathscr{D}'$ if

$$T_i(\varphi) \to T(\varphi) \quad \text{for } i \to \infty$$

uniformly on all bounded subsets B of functions φ of C_c^∞.

This definition of a distribution was given by Schwartz [1950/51]. He similarly defined the space of tempered distributions as the dual of $\mathscr{S}$, the rapidly decreasing functions. Sobolev [1936a] also used this approach.

2. Sequences. It is a fundamental theorem in the theory of distributions that *any distribution is a limit in $\mathscr{D}'$ of a sequence of continuous functions.* Thus sequences of functions give an alternative method for defining distributions. This method is very similar to Cantor's construction of the real numbers. Several sequence definitions have been given by mathematicians and physicists who claim that they are closer to physical intuition than the functional definition. What distinguishes the different sequence approaches is the definition of a fundamental sequence, which is not *a priori* given, since the space $\mathscr{D}'$ is not given in advance.

(a) A sequence f_n of continuous functions on $\mathbb{R}^n$ (or $L^1_{\text{loc}}(\mathbb{R}^n)$ functions) is said to be a fundamental sequence if

$$\int_{\mathbb{R}^n} f_i \varphi \tag{3}$$

is convergent for all $\varphi \in C_c^\infty$. Two fundamental sequences f_i and g_i are equivalent if

$$\lim_{i\to\infty} \int_{\mathbb{R}^n} f_i\varphi = \lim_{i\to\infty} \int_{\mathbb{R}^n} g_i\varphi \quad \text{for all } \varphi \in C_c^\infty. \tag{4}$$

An equivalence class of fundamental sequences is called a distribution. This sequence definition is the one which is closest to the functional definition since it makes uses of test functions as well. It was suggested by Tolhoek in 1944 (independently of Schwartz' work), by Mikusiński [1948], Lighthill [1958] and by Courant [Courant–Hilbert 1962, p. 777] (all three depending on Schwartz' work).

(b) A sequence of functions $f_i \in C(\mathbb{R})$ is called fundamental if for every compact subset K of $\mathbb{R}$ there exists a sequence $F_i \in C(\mathbb{R})$ and a natural number k such that

$$F_i^{(k)}(x) = f_i(x) \quad \text{for } x \in K \text{ and for all } i \in N$$

and

$$F_i(x) \text{ converges uniformly on } K.$$

Two fundamental sequences f_i and g_i are equivalent if for all compact subsets K of $\mathbb{R}$ the sequences F_i, G_i, mentioned above, can be chosen such that

$$\left.\begin{aligned} f_i &= F_i^{(k)} \\ g_i &= G_i^{(k)} \end{aligned}\right\} \quad \text{for the same } k$$

and

$$F_i = G_i \qquad \text{for all } i \in \mathbb{N}$$

An equivalence class of fundamental sequences is called a distribution. (The extension to more than one dimension is obvious.)

The distributions defined in this way are also included in Schwartz' distributions, since any Schwartz distribution is locally a derivative of a continuous function.

Definition 2(b) was suggested by Mikusiński and Sikorski [1957]. A slightly different approach was presented by Korevaar [1955].

When distributions are defined as equivalence classes of sequences, the imbedding of $C(\mathbb{R}^n)$ in the space of distributions and the definition of a derivative are given in an obvious way.

3. Formal derivatives of continuous functions

A Schwartz distribution is locally (i.e. on every compact subset of $\mathbb{R}^n$*) a derivative of a continuous function.*

Any Schwartz distribution $T \in \mathscr{D}'(\mathbb{R}^n)$ *can be written as a locally finite sum:*

$$T = \sum_{p_1 \cdots, p_n = 0}^{\infty} \left(\frac{\partial}{\partial x_1}\right)^{p_1} \cdots \left(\frac{\partial}{\partial x_n}\right)^{p_n} f_{p_1, \ldots, p_n}, \tag{5}$$

where $f_{p_1, \ldots, p_n}$ *are continuous functions (i.e. for any compact set* $K \subset \mathbb{R}^n$ *all but a finite number of the* f_i*s vanish on* K*).*

These theorems gave rise to the following definitions:

(a) Consider all locally finite sums of pairs

$$\sum_k (\Omega_k, f_k), \tag{6}$$

where the Ω_ks are differential operators and the f_ks are continuous functions. Two such expressions are called equivalent if the results obtained from using formal partial integration on the integrals

$$\int_{\mathbb{R}^n} \left(\sum_k \Omega_k f_k\right) \cdot \varphi \tag{7}$$

are the same for all test functions $\varphi \in C_c^\infty(\mathbb{R}^n)$.

Equivalence classes of such expressions are called distributions. This definition was given by Tolhoek [1949] and Courant [Courant–Hilbert 1962, p. 775].

(b) Consider locally finite power series

$$\sum_{p_1, \ldots, p_n = 0}^{\infty} f_{p_1, \ldots, p_n} z_1^{p_1} \cdots z_n^{p_n}, \tag{8}$$

where the $f_{p_1, \ldots, p_n}$s are continuous functions.

Two such series are equivalent if their term-by-term difference is the sum of terms of the form

$$f(x) z_1^{p_1} \cdots z_\nu^{p_\nu} \cdots z_n^{p_n} - g(x) z_1^{p_1} \cdots z_\nu^{p_\nu + 1} \cdots z_n^{p_n},$$

where $f(x) = (\partial/\partial x_\nu) g$.

An equivalence class of locally finite power series is called a distribution. The mapping which sends the power series (8) into the functional T:

$$T(\varphi) = \sum_{p_1, \ldots, p_n} (-1)^{p_1 + \cdots + p_n} \int_{\mathbb{R}^n} f_{p_1, \ldots, p_n}(x) \left(\frac{\partial}{\partial x_1}\right)^{p_1} \cdots \left(\frac{\partial}{\partial x_n}\right)^{p_n} \varphi(x)\, dx$$

is an isomorphism between the distributions defined here and those defined by Schwartz.

Definition 3(b) was advanced by H. König [1953].

(c) Consider all pairs (f, n) of a continuous function f on a fixed interval $[a, b] = K$ and a natural number n. Two such pairs (f, n) and (g, m) are equivalent if

$$I^m f - I^n g \quad \left(I^n \text{ is the } n \text{ times iterated integral } \int_{(a+b)/2}^{x} \cdot \right)$$

is of the form

$$\sum_{i=0}^{m+n-1} a_i x^i.$$

The equivalence class which contains (f, n) is denoted $[f, n]$. A system $T = \{[f_K, n_K]\}$ of equivalence classes corresponding to any compact interval $K \subset \mathbb{R}$ is called a distribution if for all $K' \subset K$ $[f_{K'}, n_{K'}]$ is a restriction of $[f_K, n_K]$ (with an obvious definition of a restriction). The extension to $\mathbb{R}^n$ is easy.

R. Sikorski [1954] and S. e Silva [1955] invented this definition.

It is clear how to imbed continuous functions in the space of distributions and how to define differentiation when the definitions in §3 are used.

4. Mikusiński's operators. Consider the ring of continuous complex-valued functions on $\mathbb{R}_+ \cup \{0\}$ with the compositions

$$(f + g)(t) = f(t) + g(t),$$

$$(f * g)(t) = \int_0^t f(u) g(t - u)\, du.$$

Since this ring has no zero divisors [Titchmarsch 1926], it can be extended to a field. Mikusiński, who discovered this generalization of the function concept [1950, 1959], called the elements in the extension field operators. He generalized the operators to operators which did not need to have "support" on a positive half-line [1959] and showed that these "distributions" were not equivalent to Schwartz' distributions (see also Lützen 1979, Ch. V).

5. Hyperfunctions. Consider complex functions which are holomorphic in $\mathbb{C} \backslash \mathbb{R}$. Two such complex functions are called equivalent if their difference is holomorphic in the whole complex plane. The space of equivalence classes $H(\mathbb{C}\backslash\mathbb{R})/H(\mathbb{C})$ is called the space of hyperfunctions. It was shown by Bremermann [1965, p. 50] that for every distribution $T \in \mathscr{D}'(\mathbb{R})$ there eixsts a complex function, holomorphic on $\mathbb{C}\backslash\text{supp}\, T$ such that

$$f(x + i\varepsilon) - f(x - i\varepsilon) \underset{\varepsilon \to 0}{\to} T \qquad \text{ind } \mathscr{D}'.$$

In this way $\mathscr{D}'$ can be imbedded in the space of hyperfunctions. To define hyperfunctions in more dimensions is considerably more difficult.

Hyperfunctions were introduced by Sato [1959/60] but had already been anticipated by several other mathematicians (see Ch. 3, note 18).

6. Nonstandard functions. Laugwitz and Schmieden [1958] and Robinson [1961] extended the field of real numbers to a ring and a field, respectively, including infinitely large and infinitely small numbers. Functions in the extended ring or field give interesting generalized functions. For example, the quasi-standard function

$$\delta(x) = \frac{\omega}{\sqrt{\pi}} e^{-\omega^2 x^2},$$

where ω is an infinite natural number ($\omega \in \mathbb{N}^*\backslash\mathbb{N}$), has the property characteristic of the Dirac δ-function:

$$\int \delta(x) f(x) = f(0),$$

in the sense that the standard parts of the two sides of the equation are equal.

In this way distributions can be represented by nonstandard functions, but the correspondence between quasi-standard functions (a subclass of the nonstandard functions) and distributions is not 1–1, since there are many quasi-standard functions representing each distribution. By forming suitable equivalence classes of quasi-standard functions, a 1–1 correspondence can be established. Equivalence relations of this kind have been defined in various ways by Laugwitz [1961], Luxemburg [1962] (similar to the method described in §2(b) of this Appendix) and Robinson [1966] (similar to Schwartz' approach). It is worth noting that multiplication of distributions can *not* be defined since it would depend on the representatives of the equivalence classes. Thus in this respect the original nonstandard functions are easier to handle.

Notes

Introduction

[1] Bourbaki, for example, in [1948] refrained from answering the philosophical question about the connection between the experimental world and the mathematical world, but he stated:

> Qu'il y ait une connection étroite entre les phénomèmes expérimentaux et les structures mathématiques, c'est ce que semblent bien confirmer de la façon la plus inattendue les découvertes récentes de la physique contemporaine; mais nous en ignorons totalement les raisons profondes ..., et nous les ignorerons peut-être toujours ...; mais d'une part la physique des quanta a montré que cette intuition "macroscopiques" du réel couvrait des phénomènes "microscopiques" d'une toute autre nature relevant de branches des mathématique que *n'avait certes pas été imaginées en vue d'applications aux sciences expérimentales.* (My italics.)

One such profound reason for the applicability of functional analysis to quantum mechanics was given in the same book by de Broglie [1948]. He indicated that it was no mystery that functional analysis could be used to describe the "mechanique ondulatoire" since its creation had been motivated by problems of vibratory motion.

[2] In his monograph also [1978] Dieudonné takes the same point of view. J. Fang [1970] used the theory of distributions to argue that Bourbaki is not a sterile mathematician:

> Is the modern theory of partial differential equations therefore necessarily and hopelessly abstract? Hardly. Even if topological vector spaces or functional analysis in general barely might be considered abstract by some, the latter would not hesitate to regard as concrete the manner in which the theory of distribution had elegantly and rigorously rationalized Dirac's delta-function in mathematical physics. In this sort of contexts, then Bourbaki can never be grouped with "sterile" and "abstract" mathematicians whose moronic existence is based on certain "vacuous" axioms.

[3] From Ch. 6 it will be seen that the theory of partial differential equations was the main object for Schwartz. The "elegant rationalization of the delta-function" was "presented in the process".

[4] An excellent, elementary, and well-motivated treatment of the theory of distributions can be found in L. Schwartz' *Méthodes Mathématiques pour les Sciences Physiques* [1961]. However some of the topological considerations are omitted in this textbook.

Chapter 1

[1] In my opinion the reason why Sobolev's work on distributions was not carried to the fruitful stage to which Schwartz carried the theory is not to be found in an insufficient knowledge of functional analysis but in the lack of sufficiently diverse motivating factors.
[2] Fantappié [1943a] begins with an interesting historical survey of the use and theories of functionals.
[3] A complex function on the complex sphere is called ultra-regular if it is locally analytic, i.e. regular in its domain of definition, and if it is 0 at the point ∞ (if this point belongs to its domain) [Fantappié 1943a, Ch. II].
[4] Let $y_0(t)$ be an ultra-regular function on a set M_0. Then a typical neighbourhood (A, σ) of y_0, corresponding to a compact set $A \subset M_0$ and a positive real number σ, consists of all functions y, ultra-regular on a set $\supset A$ for which

$$|y(t) - y_0(t)| < \sigma \quad \text{for } t \in A,$$

[Fantappié 1943a, §8]. By (A) Fantappié denoted the space of ultra-regular functions defined and analytic in an open set $\supset A$. Then $(A) = \bigcup_{\sigma \to \infty} (A, \sigma)$.
[5] If F is defined on (A) (cf. note 4) then $\gamma(\alpha)$ is defined on $B = \mathbb{C} \setminus A$.
[6] The contour C must separate the complement of the domain of the function y from the complement of the domain of the indicatrix γ (i.e. A as in note 5)

$||||$ ~ domain of y
$\equiv$ ~ domain of $\gamma = B$
$///$ ~ $A = \mathbb{C} \setminus B$

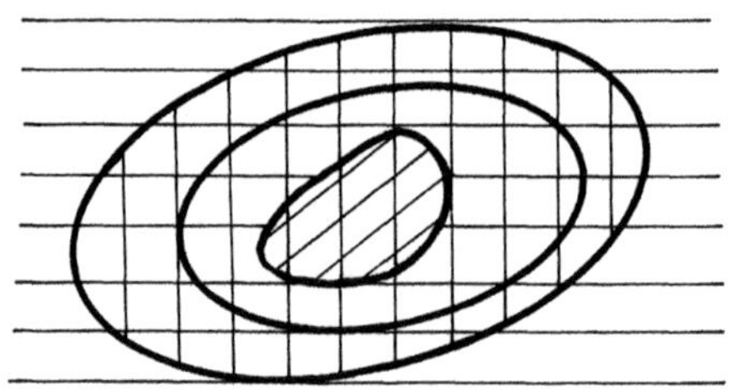

[7] These conditions are [Fantappié 1940, §41]:

(a) $(g_1 + g_2)(B) = g_1(B) + g_2(B)$.
(b) $g_1 \cdot g_2(B) = g_1 \circ g_2(B)$.
(c) If g is the constant function 1 (i.e. $g(\lambda) = 1$) then $g(B)$ is the identity operator. If g is the identity (i.e. $g(\lambda) = \lambda$) then $g(B) = B$.
(d) If $g(\lambda, \alpha)$ depends analytically on a parameter α then $g(B, \alpha)(f)$ is analytic in α for all f in the domain of B.

[8] Note that, according to (5) and (6), the indicatrix of (12) (formed as in (9) and (10)) is precisely γ.
[9] In particular it satisfies (a)–(d) in note 7.
[10] As far as I have been able to see, Fantappié's operational calculus has not been used very much for practical purposes. (See, however, Fantappié [1943b].)
[11] After Schwartz had seen that functionals on a space A (e.g. $A = \mathbb{R}$) could be used as generalized functions, Fantappié's theory suggested that the corresponding indicatrices defined on $\Omega \setminus A$ ($\mathbb{C} \setminus \mathbb{R}$) could be used as generalized functions on A as well. This led to the theory of hyperfunctions (Ch. 3, note 18).

Chapter 2

[1] This use of the term generalized solution differs from that used in potential theory (see note 31).
[2] If I had only considered those instances where the generalized derivative or solution was a generalized function the prehistory would have been reduced to only a few sections, and would not have been representative of the range of methods of which the distribution theory was a synthesis.
[3] The main lines in the following were already clear to me before the Edinburgh congress. However I have two new facts from Demidov: d'Alembert's later opinion in his *Opuscule*, Vols. 8 and 9, and Lagrange's use of test functions in [1761].
[4] According to Demidov [1977] d'Alembert applied this criterion to the wave equation in the ninth and unpublished volume of his *Opuscules*.
[5] In [1780] d'Alembert gave the following geometric argument: If $\Phi(x - y)$ at the point $x - y = A$ changes expression from ψ to φ and these two functions have different

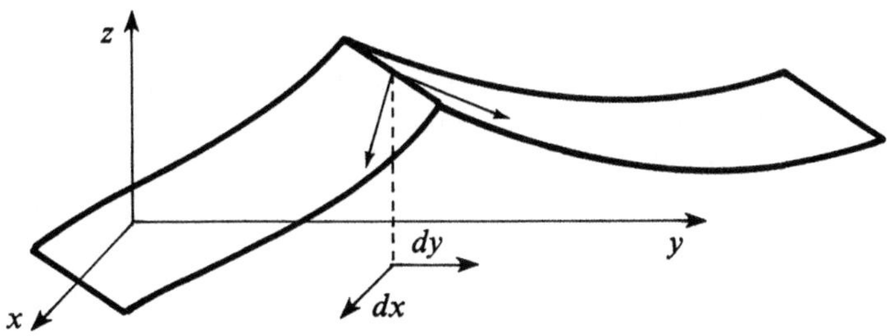

derivatives at $x - y = A$, then the tangents determined by the positive dx and dy directions at a point on the line $x - y = A$ will not span the tangent plane in this point.

This argument is strange in several respects:

(1) Where is the differential equation?
(2) Where is the inconsistency?
(3) What is the tangent plane at a point on the "ridge" of the roof?

The answers to (1) and (2) seem to be that d'Alembert felt that the graph of a solution to a partial differential equation must have a tangent plane which is spanned by the described tangents. The answer to (3) can apparently not be determined completely, but it is intuitively obvious that a tangent plane must contain the line

$$\{(x, y, z) \mid x - y = A \wedge z = \varphi(A) = \psi(A)\}.$$

However that is not the case with the plane spanned by the aforesaid lines.

The argument is probably inspired by Monge, see Taton [1950] and (note 11).
[6] It is unclear whether Euler realized that the alteration of the first derivative would change the function f itself in the whole half of its domain, but the observation does not invalidate Euler's argument since the ordinate difference between the two curves is still infinitely small.

Where the argument b operated with infinitely small quantities along the abscissa axis, argument (c) operates with infinitesimals in the direction of the ordinate.

[7] Thus a description in modern terms of Euler's ideas on the calculus is not possible within classical analysis unless extended to the theory of distributions. Another description has recently been made possible by nonstandard analysis, in which distributions are naturally described by (equivalence classes of) analytic expressions (Appendix, §6). This seems to support Robinson's idea that the history of the calculus ought to be rewritten in terms of nonstandard analysis, but again one should be careful. Mathematics from other epochs may be *compared* to modern theories as I have done here. Nevertheless, it ought to be described and understood on its own premises so as not to be translated and embedded in a modern theory. For example, it would be absurd to attribute knowledge of distributions to Euler, even if distributions are nothing but analytic expressions in nonstandard analysis (see §4, note 15). Euler could not even imagine how badly E-continuous *functions* could behave.

[8] We have seen that Euler also, to some extent, advocated the substitution of the differential equation with another procedure. There is, however, the big difference between Lagrange's and Euler's substitutions that where Euler found his substitutions by purely mathematical reasoning, Lagrange's substitution was based on a physical reinvestigation of the problem. For us, who are interested in generalized solutions, Euler's procedure is clearly the most interesting, but from a physical point of view Lagrange has the most satisfactory approach, for it is in no way clear to what extent a mathematical generalization of a differential equation continues to give a correct description of the physical reality when the assumptions under which the differential equation were originally derived no longer hold (the assumptions here being E-continuity in the eighteenth century and twice differentiability in the late nineteenth century).

[9] Lagrange did not use this notation for the definite integral but described verbally that in the integral "prise en sorte qu'elle évanouisse, lorsque $x = 0$, on fait $x = a$".

[10] Laplace considered the differential equation to be a limiting case of difference equations and found the (generalized) solutions as the limits of the solutions to the difference equations. As far as I know such a procedure was not suggested later as an explicit method for defining generalized solutions.

Laplace's method is very similar to Lagrange's first method (§11, start), but it is not based on physical but rather on purely mathematical reasoning.

[11] D'Alembert who listened to Monge advance his ideas in a talk at the Paris Academy in November 1771 would not discuss them with Monge, but as pointed out in (note 5) they probably motivated him to his geometric argument of [1780] opposing Monge's point of view [Taton 1950].

[12] Arbogast treated, for example, the equation

$$\frac{\partial z}{\partial x} = -\frac{\partial z}{\partial y} \tag{6}$$

which d'Alembert had discussed in [1780] (§8): the surface $z = \varphi(x - y)$ is constructed by drawing lines parallel to the bisector $\{x - y = 0, z = 0\}$ through the completely arbitrary curve $z = \Phi(x)$ in the (x, z) plane.

If now (x, y) runs along straight lines l_1, l_2 in the (x, y) plane parallel to the x axis and to the y axis, respectively, then it is apparent that the corresponding values of $z = \varphi(x - y)$ vary in precisely the same manner when the point runs along the one line in a positive direction and along the other in a negative direction. Arbogast claimed that this proved that $z = \varphi(x - y)$ satisfied (6) for all functions φ.

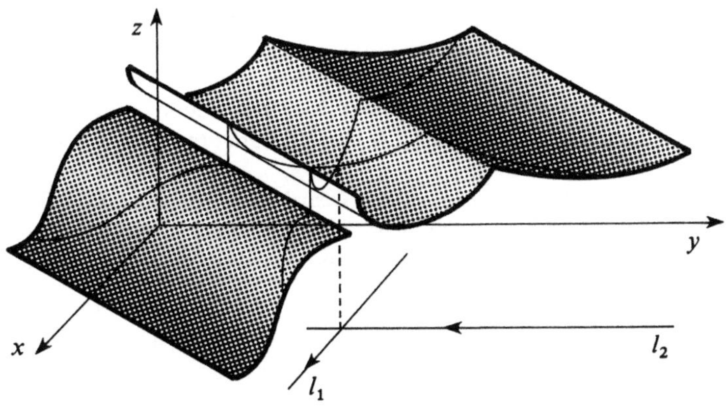

[13] The problem of the convergence of Fourier series was the one which contributed most to the rigorization program. Since Fourier series were closely related to differential equations, differential equations still influenced the foundational problems indirectly.
[14] Harnack used Fourier expansions in his description, but he saw that the propagation of the singularities could not be derived directly from the properties of the Fourier series. Instead he used Christoffel's equation to determine the *velocity* with which the singularities propagated. To determine the *amplitudes* in the singular points he used the equations

$$\int_0^x \frac{\partial^2 f}{\partial t^2} \sin nx \, dx = \int_0^x \frac{\partial^2 f}{\partial x^2} \sin nx \, dx \qquad n = 1, 2, 3, \ldots, \tag{$*$}$$

which he derived from the wave equation.

It is unclear what Harnack meant by ($*$) for in the singular points the second-order derivatives in the formulas do not exist. He probably interpreted $\partial^2 f/\partial t^2$ and $\partial^2 f/\partial x^2$ as derivatives almost everywhere in Riemann measure (see note 24) in which case ($*$) would acquire a well-defined meaning. However he did not mention this interpretation explicitly.

Harnack's method can be interpreted as a test function generalization of the wave equation with the test functions sin nx but since the derivatives are not transferred to the test functions by partial integration, the idea is even more implicit than by Lagrange [1760/61] (§11).
[15] In Riemann–Weber [1919] it is also suggested that one approximates the (generalized) solution by a finite number of terms in the Fourier expansion. (Sequence definition of a generalized solution! (see §63G)).

In [1841] Cauchy also treated discontinuous solutions physically.

As pointed out in Levy *et al.*'s biography [1967, p. 38] Hadamard also resorted to this method when dealing with discontinuities of the first order.

> Hadamard, comme les auteurs de cette époque, ne voit d'autre méthode que celle qui consiste à traiter chaque problème physique séparément, en "reprenant la mise en équation" suivant ses propres termes. Nous savons aujourd'hui que, dans un grand nombre de cas, ... les conditions que l'on obtient ainsi sont précisément celles que l'on trouve en écrivant que les équations sont satisfaites au sens de distributions, ce qui permet une discussion mathématique générale de telles discontinuités.

[16] α/β remains finite if there exists constants k, K such that $0 < k < \alpha/\beta < K < \infty$. Definition (17) is more general than the ordinary definition of the second-order derivative because in (17) the ordinary two limits have been replaced by one.
[17] I think that Riemann was aware of the fact that his definition (17) was more general than Cauchy's definition of a second-order derivative, but there is no explicit evidence to be found in Riemann's article.
[18] In his review of Schwartz' *Théorie des Distributions*, Bochner [1952] proclaims Riemann a hero in the history of the theory of distributions (actually *the* hero, next to himself, who according to his own judgement possessed the main ideas and techniques prior to Schwartz (see Ch. 3, §14 and §15)):

> And as regards the novelty of introducing "distributions" which are more general than Stieltjes integrals, say, we think that credit for it ought to be assigned to Riemann who in his paper on trigonometric series interprets a series
>
> $$\sum_{n=1}^{\infty} (\lambda_n \cos nx + \mu_n \sin nx),$$
>
> with only $\lambda_n \to 0$ and $\mu_n \to 0$ as a symbol
>
> $$\frac{d^2F}{dx^2},$$
>
> where $F(x)$ is defined as the uniformly convergent series
>
> $$-\sum \frac{\lambda_n \cos nx + \mu_n \sin nx}{n^2} \qquad (*)$$
>
> and then "convolves" the series $(*)$ with that of a testing function in the appropriate manner.

Thus Bochner gives Riemann the credit for the generalization of the concept of function with the help of test functions. This opinion requires some comments.

First of all, there is no mention of generalized functions in Riemann's work; the limiting value of (17), is only given meaning at the points where it converges. Secondly, the "symbol d^2F/dx^2" is not used by Riemann; he only speaks of the limit of (17). Thirdly, the generalized derivative is not defined in terms of test functions as Bochner indicates but by (17). "Test functions" which with Riemann are twice-differentiable functions on $[b, c]$ which vanish together with their first derivatives at b and c, are introduced not primarily to receive the differentiation, but for the purpose of localizing a certain integral [see Anmerkung 5 in Riemann's Werke]. So in reality they are not test functions but localization functions.

For these reasons I find that Bochner has given Riemann credit for something he never did. As we have seen in §11 Lagrange deserved this credit more than Riemann.
[19] Hawkins, in his book [1970] on Lebesgue, has given a fine exposition of the work done on the main theorems. Since his main interest is the definition of the integral he focuses on theorem (II). I have made extensive use of Hawkins' book for the following brief review.
[20] Cauchy defined [1823a] the integral by the procedure later used by Riemann, but restricted its domain to the continuous functions, for which he proved its existence (convergence). He did not use the terms *differentiable* and *integrable*.
[21] Riemann extended Cauchy's definition to all functions for which the mean sum involved in the definition converged.
[22] More precisely Hankel proved that $\int_0^x f(x)\,dx$ is nondifferentiable when f is Riemann's function, which is integrable but discontinuous with jumps on a dense set.

[23] For a continuous function f in an interval $[a, c]$ Dini defined the derivatives as follows: first the two auxiliary functions L_x and l_x are defined as: $(b < c)$

$$L_x = \sup_{0<h<b-x} \frac{f(x+h)-f(x)}{h}, \qquad l_x = \inf_{0<h<b-x} \frac{f(x+h)-f(x)}{h} .$$

Since L_x decreases and l_x increases with decreasing b, the following two limits exist:

$$D^+f(x) = \lim_{b\downarrow x} L_x, \qquad D_+f(x) = \lim_{b\downarrow x} l_x .$$

These two quantities, which are in modern notation the lim sup and lim inf, respectively, of the right-hand difference quotient, are two of Dini's derivatives. The other two D^-f and D_-f were defined similarly, taking supremum and infimum over intervals to the left of x. Dini proved that all four derivatives were proper generalizations of the ordinary derivative.

It is interesting to observe that as late as 1877 when Dini's note on the Dini derivatives was presented to the Accademia dei Lincei the idea that the theorems in analysis should have a sort of general validity (cf. [Lützen 1978]) was still alive. The reivewer of Dini's note in the *Atti* [Dini 1877] thus gave the following introduction: (it is unclear whether the remark stems from Dini himself)

> Posta ormai fuori di dubbio la esistenza di funzioni finite e continue in un dato intervallo che pure non hanno mai una derivata determinata e finita, restano a farsi degli studi generali pei quali vengano poste in evidenza le limitazioni da introdursi nel concetto di funzione, affinchè ad esse resti sempre applicabile il calcolo differenziale, o vengano trovati dei metodi di calcolo più generali che si applichino a qualunque funzione continua.

Translation. "Since it is now beyond doubt that in a given interval there exist finite continuous functions which have no well-determined and finite derivative, it now remains to carry out general researches by which it becomes clear which restrictions must be introduced in the concept of function to make the calculus generally applicable to these, or by which more general methods of the calculus can be found, which are applicable to any continuous function."

[24] Another generalization of theorem (I) was due to Harnack [1882]:

> Lehrsatz 8: Der Differentialquotient des bestimmten Integrales
>
> $$F(x) = \int_a^x f(x)\, dx$$
>
> ist vorwärts genommen in allgemein gleich $f(x)$. Die Stellen, an denen er von diesem Werthe um mehr als eine beliebig kleine Grösse δ abweicht oder falls $f(x)$ unbestimmt wird, die Stellen an denen die Differenz zwischen den Unbestimmtheitsgrenzen und dem Differentialquotienten grösser wird als δ oder endlich die Stellen an denen die Unbestimmtheitsgrenzen des Differentialquotienten von denen der Funktions $f(x)$ um mehr als δ differieren, bilden eine discrete Menge.

If we disregard the multiple values, the theorem asserts:

$$\forall\delta > 0, \qquad \exists \text{ a discrete set } A_\delta\ \forall x \notin A_\delta\ \exists\rho,$$

$$|h| < \rho \Rightarrow \left| \frac{F(x+h)-F(x)}{h} - f(x) \right| < \delta.$$

Harnak "proved", moreover, that if f is continuous and if $f'(x) = 0$ "in general" (im allgemein) then f is constant.

Harnack's work is interesting because it was an attempt to define differentiation a.e., a generalization of the notion of differentiation which in Lebesgue's theory proved most fruitful. Harnack's generalization, in contrast to Dini's, however, was soon abandoned, possibly as a result of the counterexamples to the theorem mentioned last, which Cantor [1884, p. 385] and Scheeffer [1884, pp. 61–68] found only two years after the theorem had been announced.

[25] The symmetric form of (II) is invalid for the Riemann integral, even when differentiable means differentiable everywhere, for as Volterra proved in [1881] there exists a differentiable function with bounded nonintegrable derivative.

[26] This is to say: an absolutely continuous function F has the derivative f in $[a, b]$ if and only if

$$F(x) - F(a) = \int_a^x f(\xi)\, d\xi \qquad \forall x \in [a, b].$$

In this formulation this generalization of the derivative comes close to a test function definition (see §63F) for test functions of the form $\chi[0, x]$ (the characteristic function on $[0, x]$).

[27] Vitali defined absolute continuity as follows. The increase of a function F over an interval $[a, b] \times [c, d]$ is defined as

$$F(a, c) + F(b, d) - F(a, d) - F(b, c)$$

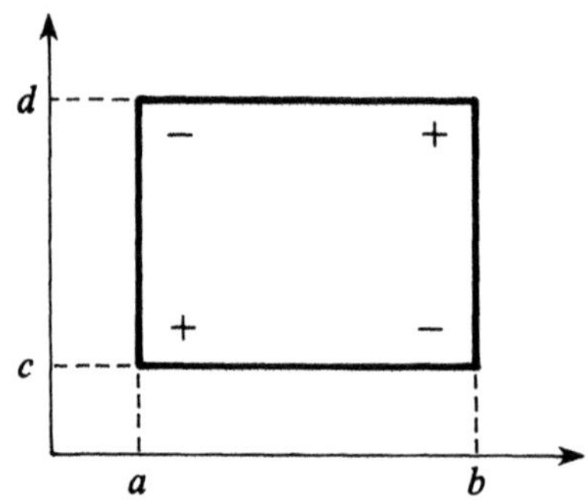

A function F defined in $R_0 = [A, B] \times [C, D]$ is called absolutely continuous if the sum of the increase of F in a countable sequence of disjoint rectangles $R_n \subset R_0$ tends to zero when the area of $\bigcup_{n=1}^{\infty} R_n$ tends to zero.

[28] In a footnote Tonelli [1926a, p. 634; 1926b, p. 1199] remarked that neither Vitali's nor his own definition of absolute continuity contained the other as a special case.

[29] The Sobolev space H_1^1 is the space of functions which have generalized partial derivatives (in the distribution sense) of the first order and which together with their partial derivatives are in L^1. According to Fubini–Tonelli's theorem the second assumption in Tonelli's definition of absolutely continuous functions implies that the integral $\int_0^1 \int_0^1 |(\partial/\partial x) f(x, y)\, dx\, dy| = \int_Q |(\partial/\partial x) f(x, y)\, dx\, dy|$ exists, i.e. that $(\partial/\partial x) f$ is in L^1. Similarly for $(\partial/\partial y) f$. According to Schwartz [1950, p. 58, Theorem 5] this is equivalent to the existence in the distribution sense of $(\partial/\partial x) f$ and $(\partial/\partial y) f$ as L^1 functions. Therefore Tonelli's space is precisely equal to H_1^1, if the requirement that u be continuous is disregarded.

[30] Discussed already in Dirichlet's lectures 1856/57 in Göttingen.
[31] During the period when the Dirichlet principle was considered invalid, H. A. Schwartz, C. Neumann and H. Poincaré devised methods to *construct* the solution u to the Dirichlet problem from knowledge of the boundary value. These constructions would lead to the solution if such a solution existed and would otherwise give a harmonic function with wrong boundary values. Such an operator which maps a pair (Ω, f) consisting of an open set and a boundary value f on $\partial\Omega$ onto a harmonic function, solving the corresponding Dirichlet problem if a solution exists, is called a generalized solution to the Dirichlet problem. This is a generalization of another kind than those we are interested in. It relaxes the boundary conditions whereas we are looking for methods relaxing the differentiability conditions.
[32] For the history of the Dirichlet principle the reader is referred to Anger [1961], Brelot [1972], and Monna [1975]. There are copious bibliographies in all three.
[33] The most significant aspect of the further development was thus connected to the conditions on the boundary curves. However, in order not to overburden the account with too many technicalities, I shall leave out the precise description of the admissible boundaries. Another problem I shall omit is how to interpret the phrase "u takes the values f on the boundary $\partial\Omega$". This expression loses its obvious meaning when u is not assumed continuous, but as discussed by Courant and Hilbert [1937, p. 482] it can be given a perfectly unambiguous meaning even in the discontinuous case.
[34] Condition (5) is stated in a footnote. In the same footnote it is pointed out that in this case the existence of the Dini derivatives is equivalent to differentiability almost everywhere.

Beppo Levi pointed out that even though the conditions (3a) and (3b) might seem unnatural, they are in fact necessary for the solution of the variational problem; for, as he showed [Levi 1906, §6–8 and note, §50], without this condition the Dirichlet integral can approach zero as closely as one wishes without attaining this minimum.
[35] We will see in the next chapter that this is a characteristic feature of elliptic equations such as Laplace's equation. All generalizations which have been proposed for this equation (also in its variational formulation as we are dealing with here) have proved to have precisely the same solutions as the classical equation in C^2.
[36] The only condition which is not clearly fulfilled in Tonelli's definition is (2°), i.e. that the total variation of $u(\bar{x}, y)$ in $[0, 1]$ is integrable in $\bar{x} \in [0, 1]$. Now the total variation of $u(\bar{x}, y)$ is given by

$$V_y(\bar{x}) = \int_0^1 \left| \frac{\partial u}{\partial y}(\bar{x}, y) \right| dy \quad \text{a.e.}$$

(exists according to (3b)), thus we want to prove the integrability of this function, which according to Fubini–Tonelli's theorem amounts to proving that $\partial u/\partial y$ is integrable in the square $Q = [0, 1] \times [0, 1]$. This follows from Levi's condition (5) which implies that $\partial u/\partial y$ is in $L^2(Q)$ and thus in $L^1(Q)$.
[37] In this connection Tonelli extended the definition of absolute continuity from a square to any open bounded set. In 1929 it was known that the crucial point in the application of direct methods in variational calculus was the semicontinuity of the variational integral in the space of admissible functions. Tonelli proved this semicontinuity.

In the case of the Dirichlet integral Tonelli's admissible functions are identical with Beppo Levi's.

[38] In an earlier paper [1933b] Nikodym had shown the use of his method in the special case of the Dirichlet problem of the Laplace equation. Let me briefly discuss the main ideas for this case, thus indicating how Beppo Levi's spaces became important to Nikodym.

(I) For a sufficiently nice function φ defined in $\bar{\Omega}$ we seek a harmonic function ψ on the domain Ω which has the same values as φ on $\partial\Omega$. That is, we wished to split φ into a sum

$$\varphi = \psi + \eta \quad \text{where } \psi \text{ is harmonic in } \Omega \quad \text{and} \quad \eta = 0 \text{ on } \partial\Omega. \tag{$*$}$$

Nikodym, however, did not try to solve the Dirichlet problem (I), but the following "transformed Dirichlet problem" (II), which Zaremba had connected to problem (I) in [1909].

(II) To a sufficiently nice function φ in $\bar{\Omega}$ find a harmonic function ψ in Ω such that

$$\int_\Omega (\operatorname{grad} \varphi - \operatorname{grad} \psi) \cdot \operatorname{grad} h \, dv = 0 \tag{$**$}$$

for all harmonic functions h in Ω.

Zaremba could easily prove that if a solution to this transformed problem existed, it was uniquely determined except for a constant (to see that two solutions ψ_1, ψ_2 differ by a constant use $h = \psi_1 - \psi_2$).

He then showed that if the Dirichlet problem (I) has a solution for an open bounded domain Ω and φ sufficiently nice [Zaremba 1909, §14] then it satisfies the transformed problem (II). This is easily seen if φ and Ω are regular enough for Green's formula

$$\iint_{\partial\Omega} \eta \frac{\partial h}{\partial n} \, ds = \iiint_\Omega (\eta \Delta h + \operatorname{grad} \eta \operatorname{grad} h) \, dv \tag{$***$}$$

to hold, by taking $\eta = \varphi - \psi$ and h harmonic in ($***$). Zaremba's proof of the connection between the Dirichlet problem and the transformed problem was much more complicated, but applicable to a more general class of domains.

Zaremba thus became interested in the existence of a solution to problem (II) (although this does not guarantee the existence of a solution to the Dirichlet problem (I)). He gave such an existence proof in [1909] and another in [1927] in the more general case in which grad φ is replaced by an arbitrary L^2 vector V and Ω need not be bounded. Zaremba's proof is very technical, establishing first the existence in a circle and then extending the result to the more general Ωs.

Problem (II) was reinvestigated by Nikodym [1933b]. "Le but de mon travail est la simplification de la démonstration de M. Zaremba."

> Je l'ai obtenue grâce à des méthodes bien connues de l'analyse moderne; empruntées à la théorie abstraite des ensembles et au calcul fonctionnel abstrait.... On verra que le théorème de M. Zaremba est une conséquence d'une lemme très simple et presque évident concernant les vecteurs abstrait. [Nikodym 1933b, p. 96.]

The mentioned lemma states the existence of the projection of a point $\bar{a}$ in a Hausdorff pre-Hilbert space A on a complete subspace B. Nikodym proved this lemma (apparently independently of the proof given by von Neumann [1930]) and showed that by taking A to consist of vector functions in $L^2(\Omega)$, B to be the space of L^2 gradients of harmonic functions (a space he proved to be complete) and $\bar{a}$ to be $V(\operatorname{grad} \varphi)$, the abstract theorem would yield the desired existence theorem (II).

In this proof and in the more general theorems in [Nikodym 1935] functions with gradients (in some generalized sense) in L^2 play a certain role. However it is worth while remarking that these concrete theorems are not merely easy applications of the abstract theory of BL spaces developed by Nikodym in [1933a]. For example the completeness of B above does not follow from the completeness of the BL spaces.

[39] In fact Nikodym defined BL functions for any open bounded set D, but I shall omit the difficulties which arose in this connection.

[40] For sets D, which are as general as Nikodym assumes (note 39), he is forced to take nice subsets of D in order to secure the validity of this theorem.

[41] In [1930] Rellich proved that a sequence $u_n(x, y, z)$ of $C^1(G)$ functions, for which the integrals

$$\iiint_G u_n^2, \quad \iiint_G \frac{\partial u_n}{\partial x}, \quad \iiint_G \frac{\partial u_n}{\partial y}, \quad \iiint_G \frac{\partial u_n}{\partial z}$$

are bounded, has a subsequence that converges in $L^2(G)$. Or in modern terms: a subset of $H_2^1(G) \cap C^1(G)$ which is bounded in the Sobolev space $H_2^1(G)$ is compact in $L^2(G)$. This theorem is more directly applicable to the solution of the Dirichlet principle than is Nikodym's theorem. However it is weaker in the sense that it requires the functions to be C^1.

[42] A modern and more general version of the BL spaces with applications to the solution of the Dirichlet problem can be found in J. Deny and J. L. Lion's "Les espaces du type de Beppo Levi" [1953/54].

[43] During the nineteenth century the connection between the calculus of variations and the theory of differential equations had implicitly introduced a way of generalizing solutions to differential equations.

The connection between variational problems and partial differential equations is given by Euler's equations. These are usually derived from the variational problem

$$\delta \int_a^b F(x, y, y')\, dx = 0 \tag{1*}$$

by substituting for the desired extremal function y another function $y + \varepsilon\lambda$, obtaining the condition

$$\int_a^b F(x, y + \varepsilon\lambda, y' + \varepsilon\lambda')\, dx - \int_a^b F(x, y, y')\, dx \gtrless 0. \tag{2*}$$

A series expansion of F in powers of ε gives

$$\int_a^b \left[\varepsilon\left(\lambda \frac{\partial F}{\partial y} + \lambda' \frac{\partial F}{\partial y'}\right) + \varepsilon^2(\cdots) + \cdots\right] dx \gtrless 0. \tag{3*}$$

This inequality is supposed to hold for all ε, which can only be the case if

$$\int_a^b \left[\lambda \frac{\partial F}{\partial y} + \lambda' \frac{\partial F}{\partial y'}\right] dx = 0. \tag{4*}$$

In the simplest case of a variational problem with prescribed fixed values for y at the endpoints a, b, the function λ has to vanish in these points and thus partial integration of the last term of (4*) yields

$$\int_a^b \lambda\left(\frac{\partial F}{\partial y} - \frac{d}{dx}\frac{\partial F}{\partial y'}\right) dx = 0. \tag{5*}$$

and since this is supposed to hold for all λ (of some general class of functions)

$$\frac{\partial F}{\partial y} - \frac{d}{dx}\frac{\partial F}{\partial y'} = 0. \qquad (6^*)$$

This chain of arguments leading from the variational problem to Euler's equation was carefully studied by Du Bois-Reymond [1879] in connection with the problem of the shortest curve between two points. In particular he proved the validity of the step from (5*) to (6*) if λ is allowed to be any infinitely often differentiable function and the expression in (6*) is assumed continuous. In his proof examples of C_c^∞ functions were used.

The step from (4*) to (5*) is the most interesting to us. Du Bois-Reymond remarked [1879, §7] that in (4*) $\partial F/\partial y'$ need only be integrable whereas in (6*) it must be differentiable with an integrable derivative. He saw that even if the last assumption were not fulfilled the equation (4*) could (at least in his special case) be solved directly without the detour around (5*).

These reflections show that Du Bois-Reymond saw that (4*) was a generalization of (5*), but since the argument in the calculus of variations usually went from (4*) to (5*) such a procedure for actually generalizing derivatives ($d/dx\ \partial F/dy'$) was not explicitly realized and applied. However the method is interesting in that it is a very early anticipation of the test function generalization of differentiation.

The discussion in this note suggests a way of generalizing a differential equation: find a variational problem for which the given differential equation is the Euler equation. Then the variational problem will not involve derivatives of as high an order as the differential equation. The solution of the variational problem can then be said to be a generalized solution of the differential equation. Such a procedure for transforming differential equations into variational problems was well known at the end of the nineteenth century (cf. the Dirichlet principle), but I do not know of any examples in which the method was used explicitly to generalize the solutions.

[43a] An historical account of potential theory before 1900 can be found in Burkhardt and Meyer [1900].

[44] The Newtonian potential is expressed by

$$V(\bar{r}) = \int \frac{\rho(\bar{r}')}{|\bar{r} - \bar{r}'|}\, d\bar{r}' \qquad \bar{r} = (x, y, z);\ \bar{r}' = (x', y', z').$$

[45] Petrini wrote that θ was a "fonction à intégrale nul" in the sense that

$$\int_{T'} \theta\, d\tau = 0$$

for all domains T'. It is not stated explicitly which integral or how irregular the domains T' were that he had in mind, but it appears implicitly that he used the Riemann integral.

[46] The treatment of the propagation of singularities in the nineteenth century which I discussed in §14 and §15 was also based upon a substitution of the differential equation (the wave equation) by another equation. However the procedure in the nineteenth century differed from the methods used in the twentieth century in two respects:

(a) In the nineteenth century the substituted equations were not equivalent to the original equation even for sufficiently regular functions. They only governed the propagation of the singularity.

(b) In the nineteenth century the new equation was found by returning to the physical system and setting up new laws for it. In the twentieth century mathematical considerations replaced the physical ones, at least in the cases of interest to us.

[47] Evans [1920, p. 254] remarked: "The operator (39) has been considered somewhat roughly by Ignatowsky [1909/10] where in three dimensions it occurs as a vector function . . . , but his treatment is not exact at all points." In fact Ignatowsky's treatment of vector analysis was very physically oriented. Thus it was never mentioned how regular the functions in the definitions and the theorems must be, and hence it was not apparent that the definitions in the book (e.g. (39)) were more general than the ordinary definitions. I doubt that the author was aware of this point.

[48] More precisely:

$$F(S) = f(\sigma) + \tfrac{1}{2}f(S') + \textstyle\sum \theta_i f(M^i),$$

where σ is the open set bounded by the curve S, S' consists of all the points of S except for the corners, M_i ($i = 1, 2, 3, \ldots, n$) denotes the finitely many corners of S, and θ_i is a measure of the angle between the half-tangents at the corner i.

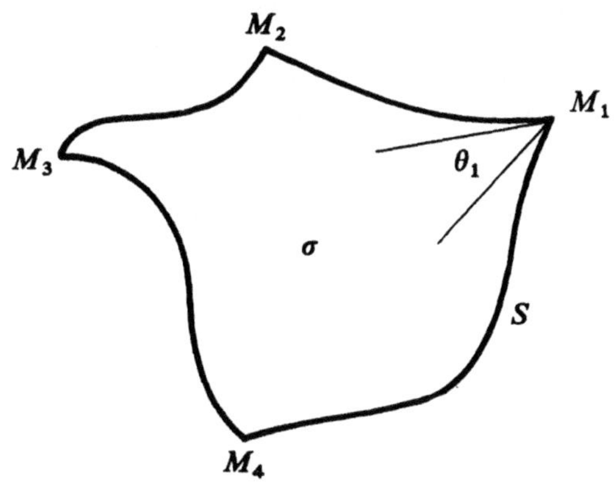

[49] The theory had the disadvantage that it was developed in $\mathbb{R}^2$ whereas the most interesting examples are in $\mathbb{R}^3$. Later on the Stieltjes integrals were also applied in three- or higher-dimensional potential theory (cf. [Anger 1961]).

In connection with the Laplace equation Evans [1920] proved the following "extension of a well-known theorem of Bôcher";

Theorem. If $u(M)$ is a potential function for its gradient vector ∇u [in the generalized sense] and if the equation

$$\int_S \nabla_n u \, ds = 0 \qquad (*)$$

is satisfied for every S of Γ in Σ, then the function $u(M)$ has merely unnecessary discontinuities, and when these are removed by changing the value of $u(M)$ at most in the points of superficial measure zero the resulting function has continuous derivatives of all orders and satisfies Laplace's equation

$$\frac{\partial^2 u}{\partial x^2} + \frac{\partial^2 u}{\partial y^2} = 0 \qquad (**)$$

at every point.

In [1928] Evans extended the theorem even further, assuming the identity (*) to hold only for almost all rectangles, i.e. rectangles "formed from lines $x = a$, $y = b$, except possibly those which correspond to values of a and b constituting a set of zero measure". In the proof [1928] he first showed that the function

$$u_\mu(M) = (\tfrac{1}{4}\mu^2) \int_{-\mu}^{\mu} \int_{-\mu}^{\mu} u(x + \xi, y + \eta)\, d\xi\, d\eta$$

satisfied the assumptions in Bôcher's theorem and therefore was a harmonic function. The theorem could then be obtained by passing to the limit $\mu = 0$.

[50] Inspired by the applications of absolutely continuous functions to the calculus of variations, Tonelli proved in [1928/29] that a function which is absolutely continuous in Cartesian coordinates is also absolutely continuous in polar coordinates.

[51] Morrey was obviously also inspired by a third discipline. As was mentioned in §20, he had already in [1933] extended Tonelli's theorem on areas of surfaces using a class of functions closely related to Tonelli's absolutely continuous functions.

[52] In the proof Calkin used the mean value procedure described in (note 49).

[53] In Calkin and Morrey's 1940 paper one looks in vain for references to Sobolev and Friedrichs. As we shall see in §60 Sobolev in 1938 developed a theory of what are now called Sobolev spaces and which are identical with the $\mathfrak{B}_\alpha$ spaces except for a different definition of the generalized derivatives. Within the theory of differential operators on Hilbert spaces, which especially interested Calkin, Friedrichs had developed similar ideas as early as 1934. The missing reference to Friedrichs, whose work was very well known in Germany, is puzzling since Friedrichs had fled from the Nazis to the United States in 1937.

In [1964] Morrey called his spaces Sobolev spaces.

[54] Weyl later specified that v must be of class $\Gamma = C_c^1(G)$.

[55] This is a generalization of Zaremba's result in the following sense.

Weyl proved that locally (i.e. in every cube included in G) $\mathfrak{E}$ consists of functions of the form grad η where η is harmonic. Thus (63) says, roughly speaking, that a gradient vector ($\in \mathfrak{F}$) can be split into an orthogonal sum (in L^2) of a gradient of a harmonic function ($\in \mathfrak{E}$) and the gradient of a function which vanishes on the boundary ∂G of G ($\in \mathfrak{G}$). This is precisely Zaremba–Nikodym's theorem.

Weyl also split other function spaces into orthogonal subspaces.

[56] My reasons for believing that Weyl's test function definition was not given in his lectures are the following:

(1) The test function definitions themselves are not given in the section on generalized vector analysis, and the implications (71) and (72) stand alone in a chapter which is otherwise devoted to the starred operators.

(2) The test function definition of the irrotational and solenoidal vectors is so connected with the proof of Weyl's version of Zaremba's theorem that it is likely that it was created in connection with this theorem.

(3) Weyl wrote in [1940]: "I depend above all, on two papers by K. Friedrichs" and he explicitly refers to [Friedrichs 1939]. This indicates that he received the idea of using test functions from Friedrichs, who used it in his [1939] paper.

(4) In his introduction (cited in §38) Weyl first suggested a generalization of the equation rot $f = 0$ using test curves, but had to reject this procedure since the integral along curves was not assumed to exist. Thus in 1940 he found a test curve definition to be the most natural. The reason could be that he had only worked with the closely related test surface definition before and not with the test function definition.

[57] The theorem is due to Riesz (1930), but Radó obtained his proof from Evans [1935, p. 237]. Evans' proof only used ordinary differentiation.
[58] In connection with theorem (74) and Radó's remark on it, Brelot [1972] wrote: "Les distributions de Schwartz ont permis plus tard d'étendre les démonstrations élémentaires."
[59] For a physicist or an applied mathematician the questions concerning irregular cases are often the natural ones, for which reason they sometimes try to solve them before the regular cases have been studied thoroughly. We have seen such treatments in §14.
[60] Here Wiener refers to Bôcher [1905/06] and Evans [1914].
[61] Wiener's confusion which was noted by Freudenthal in his biography on Wiener in the *D.S.B.* might offer an additional explanation for the missing link between the two definitions.
[61a] This quote is the continuation of the quote at the beginning of §45.
[62] This converse theorem stated that if a sequence u_n of functions in $C^1(\mathbb{R}^3) \cap L^2(\mathbb{R}^3)$ has uniformly bounded $L^2(\mathbb{R}^3)$ norms, derivatives $\partial u_n/\partial x_i$ in $L^2(\mathbb{R}^3)$ and the derivatives converge weakly in $L^2(\mathbb{R}^3)$ to U_i $(i = 1, 2, 3)$, then there exists a function $U \in L^2(\mathbb{R}^3)$ such that u_n converges to U strongly in $L^2(\omega)$ for all bounded subsets ω of $\mathbb{R}^3$, and the U_is are the quasi-derivatives of U.

Leray also proved that the quasi-derivatives were unique if they existed.
[63] In order to state Leray's result precisely we need yet another generalization.

Leray defined the quasi-divergence of a vector function $U_i(x) \in L^2(\mathbb{R}^3)$ as the $L^2(\mathbb{R}^3)$ function $\theta(y)$ (if it exists) which satisfies

$$\iiint_{\mathbb{R}^3} \left[\sum_i U_i(y) \frac{\partial a}{\partial y_i} + \theta(y) a(y) \right] dy = 0$$

for all functions a in $C^1(\mathbb{R}^3) \cap L^2(\mathbb{R}^3)$ with derivatives in $L^2(\mathbb{R}^3)$. They Leray called a solution $U_i(x, t)$ to (87) (in the generalized sense) turbulent if it was in $L^2(\mathbb{R}^3)$, had quasi-divergence 0, had quasi-derivatives $U_{i,k}(x, t)$ for a.a. $t > 0$ and satisfied certain inequalities, the exact form of which are unimportant here. His main theorem was:

> **Théorème d'existence.** Supposons donné à l'instant initial un état initial $U_i(x)$ tel que les fonctions $u_i(x)$ soient de carrés sommables sur π et que le vecteur de composantes $U_i(x)$ possède une quasi-divergence nul. Il correspond à cet état initial au moins une solution turbulente, qui est définie pour toutes les valeurs du temps postérieurs à l'instant initial. [Leray 1934, p. 241.]

In addition, Leray proved that outside of the null set where U_{ik} did not exist, the turbulent solution was an ordinary solution to the Navier–Stokes' equation.
[64] Lewis knew Bôcher's and Evans' [1914] test curve generalizations, but they were not general enough for his purpose since they assumed the existence of the first derivatives everywhere.
[65] Note that a weak solution of class C^N need not be a C^N function.
[66] In the proof of the converse theorem Bochner used the same averaging procedure as Evans [1928] (note 49). Bochner referred to Calkin and Morrey [1940] "for the role of h-averages in the calculus of variations".
[67] This is probably the first use of the term test(ing) function. Bochner has reportedly claimed that he was the inventor of the phrase testing function. In [Bochner and Martin 1948] testing functions were defined to be C_c^∞ functions.
[68] Having proved the equivalence of (90) and (92) for regular functions, Bochner and Martin [1948, p. 160] continued.

> This suggests, however, a generalization of the concept of a solution of (90). We will not push the generalization to the extreme limit possible. We will require once for all that $f(x)$ shall be defined and measurable in D except for a set of measure zero, and that in every closed subcube R, $f(x)$ shall be Lebesgue integrable. We do not require that $f(x)$ have a finite Lebesgue integral $\int |f(x)|\, dv_x$ in all of D. On the other hand we do require that $f(x)$ shall be a Lebesgue measurable point function, although we could replace $f(x)$ by a distribution, that is a set function $F(A)$.

This is a remarkable statement which shows that the authors were aware of the possibilities of further generalizations to the space of measures. The influence of Schwartz' theory of distributions, the main ideas of which had been known for three years, cannot be excluded: indeed such influence seems probable.

[69] Riemann proved it only in two dimensions.

[70] For this "principally important" extension Courant and Hilbert referred to a "demnächst erscheinende Abhandlung von K. Friedrichs zur Anwendung der Allgemeinen Operationstheorie auf Differentialoperatoren". They undoubtedly referred to Friedrichs [1939] which I shall discuss in the next section.

[70a] The early axiomatic definitions of Hilbert spaces always contained a separability axiom. In the following I shall follow this custom and assume separability of the Hilbert spaces.

[71] Let A be an (unbounded) operator defined in the domain $D(A)$ which is dense in the Hilbert space H. The domain $D(A^*)$ of the adjoint operator A^* consists of all g in H for which

$$(Af, g)$$

is a continuous function of f in $D(A)$. For $g \in D(A^*)$ the image $A^*(g)$ is defined to be the uniquely determined point in H such that

$$(Af, g) = (f, A^*g) \qquad \forall f \in D(T), g \in D(T^*).$$

A is called self-adjoint if it is densely defined and $A = A^*$, which implies that $D(A) = D(A^*)$.

A is called symmetric if

$$(Af, g) = (f, Ag) \qquad \forall f, g \in D(T).$$

which implies that A^* is an extension of A.

As an example of a differential operator on $L^2(\mathbb{R}^3)$ take the Schrödinger operator $\Delta - v$ where v is a sufficiently "nice" real function. It is defined in its natural domain, the Sobolev space $H_2^2(\mathbb{R}^3)$ (cf. §60), where it is symmetric. However, it is not self-adjoint, since its adjoint is defined in a larger domain than $H_2^2(\mathbb{R}^3)$.

[72] A symmetric operator A on H is semibounded if there exists a real constant γ such that

$$(f, Af) \geq \gamma(f, f) \qquad \text{(bounded below)}$$

or

$$(f, Af) \leq \gamma(f, f) \qquad \text{(bounded above).}$$

Friedrichs assumes, without loss of generality that A is positively bounded below, i.e. $\gamma > 0$.

[72a] Another such extension had already been given by Stone in [1932].

[73] "Other writers" refers, in addition to Friedrichs, to Murray who in [1935] treated the second-order differential operators in $\mathbb{R}^2$. Murray used the alternative inner product in connection with a generalization of differential operators to a space very similar to Nikodym's space of type Beppo Levi, but differing from this in assuming the existence of the second-order derivatives in a generalized sense [Murray 1935, p. 318, Definition II]. Murray referred to Nikodym's treatment of Zaremba's theorem [Nikodym 1933b] but surprisingly enough not to his introduction of the BL spaces in [1933a].

[74] Friedrichs concluded his paper [1939] by discussing the regularity of functions in the domains of the generalized differential operators. The main theorem stated that if

$$D(u), D^*Du, \ldots, \underbrace{\cdots D^*DD^*D}_{r \text{ factors}}\, u$$

are all in $L^2(G)$ then $u \in C^{r-m}(G)$, where m is defined from the dimension n of the space by

$$m = \left[\frac{n}{2}\right] - 1 \qquad [\ ] \text{ denotes the greatest integer function.}$$

Friedrichs referred to Sobolev [1936b] (see §60, §61).

[75] An examination of *Mathematical Reviews* shows that Calkin's publications stopped abruptly in 1941. As von Neumann's assistant he was involved in secret military research, and joined the Los Alamos project during the winter of 1943/44 [Ulam 1976, pp. 145–146, 169, 177, 190, 209].

[76] For a more comprehensive biography and bibliography see Ljusternik and Visik [1959].

[77] Sobolev did not use the term support. He said: "à chaque fonction φ corresponde un certain domaine borné V_φ à l'extérieur duquel la fonction φ s'annule". For convenience I shall continue to use the term "support" in the following.

[78] Already in [1933] Sobolev treated discontinuous solutions to partial differential equations in connection with the description of vibrations of a half-plane. His approach here was to treat the discontinuities, which were confined to certain surfaces, separately, setting up special equations, of a physically intuitive origin, to govern the discontinuities of the function. This approach is similar to the one used by several other rigorists (see §63A).

[79] Theorem I states that $L_\nu(A)$ is bounded in $C^0(D)$ for $\nu = [n/2] + 1$. But $L_\nu(A)$ is precisely

$$C^\nu(D) \cap B_\nu(\sqrt{A}),$$

where $B_\nu(\sqrt{A})$ is the ball of radius $\sqrt{A}$ in the Sobolev space $L_2^{(\nu)}$ and $C^\nu(D)$ is the continuous functions on D with continuous derivatives of order $\leq \nu$ in the interior of D. According to theorem I the inclusion

$$C^\nu(D) \cap L_2^{(\nu)} \hookrightarrow C^0(D)$$

is continuous. Since $C^\nu(D) \cap L_2^{(\nu)}$ is dense in $L_2^{(\nu)}$ (a proof can be found in Nečas [1967, p. 67]) the conclusion can be extended to a continuous inclusion:

$$L_2^{(\nu)} \hookrightarrow C^0(D) \quad \text{for } \nu = \left[\frac{n}{2}\right] + 1.$$

This gives theorem A for $p = 2$ and $\nu = [n/2] + 1$. If $\varphi \in L_2^{(\nu)}$ for $\nu > [n/2] + 1$ then $\varphi^{(\nu - [n/2] - 1)} \in L_2^{([n/2]+1)}$. According to the above $\varphi^{(\nu - [n/2] - 1)} \in C^0(D)$, whence

$$\varphi \in C^{\nu - [n/2] - 1}$$

This gives theorem A for $p = 2$.

80 More specifically he referred to Schauder [1935] and in [1936c] to Friedrichs [1927].
81 I do not know when or by whom these generalizations were made.
82 Young's generalized curves gave a generalization along very different lines from those considered here.

Another generalization of differential operators was suggested in [1949] by Tolhoek.

Chapter 3

1 The "appareil mathématique" invented for the solution of problems in the theory of Fourier *series* consisted of methods for the summation of divergent series. However, according to Schwartz, this apparatus did not give satisfactory solutions to the problems since one always had to make a distinction between Fourier series and trigonometric series. In the theory of distributions this distinction disappears.

Since the summation methods for divergent series did not anticipate the theory of distributions, and since the distinction between Fourier series and trigonometric series was probably regarded as an unavoidable fact and not as a problem before Schwartz, I shall omit a discussion of the Fourier *series* and concentrate on the Fourier *transforms*.
2 The constant 1 in the integrand is introduced in order to make the integral convergent at 0. Any continuous function of value 1 at zero will suffice as long as the convergence at infinity is not disturbed. The different choices will only affect Ψ by a constant term which will vanish in the inversion formula (7).
3 Hahn pointed out that the Fourier–Stieltjes integral included both the ordinary theory of Fourier integrals and the theory of Fourier series.
4 Wiener [1925] mentions Hahn's paper. However it is most probable that Wiener developed his generalized Fourier transformation independently of Hahn since the details (convergence, etc.) differed from Hahn's method and since the generalized harmonic analysis described in Wiener [1925] had a much wider scope than Hahn's theory and had a different motivation. Burkill's [1926a, b] development of the $\mathscr{F}_i$ transform depended on Wiener's work.
5 Cf. note 2.
6 Wiener did not explicitly mention the physical motivation of his work in his first article [1925] on generalized Fourier transforms. However, since the bulk of the theory of generalized harmonic analysis was contained in this paper, it is clear that Wiener already had the physical problems in mind then (see further [Levinson *et al.*, 1966]).
7 Here the appropriate polynomials of degree 1 have been introduced in order to secure the convergence at 0. Again they do not affect the final formula (16) in which Φ and Ψ are roughly speaking, differentiated twice.
8 Hahn did not mention that f must be in $L^1_{\mathrm{loc}}(\mathbb{R})$, a condition he evidently used.
9 Wiener [1926a, p. 100] remarked that the theory from [1925] had a "formal niche for functions with continuous spectra. To the present time, the author has been unable to produce a single example of a well-behaved function [i.e. satisfying (12)] with a continuous spectrum, or to demonstrate that no such function exist. In the theory here expounded [where only (20) need to be satisfied], the necessary existence proofs may be filled in with the aid of the theory of probability." Thus the further generalization gave a useful extension of the theory.

[10] Wiener gave another generalization of the Fourier transform to the functions growing slower than a polynomial at infinity. In his paper on operational calculus [1926b] he considered a function $\Phi \in C_c^\infty[0, 1]$ with $\Phi^{(n)}(1) = 0$ and $\Phi^{(n)}(0) = 1$ for all $n = 0, 1, 2, \ldots$. He defined

$$\Psi(\xi, \lambda) = \frac{\sin \alpha\xi}{\pi\xi} + \frac{1}{\pi}\int_0^1 \Phi(u) \cos(u + \lambda)\, du$$

and showed that the integral

$$f_\lambda(x) = \int_{-\infty}^{\infty} f(\xi)\Psi(\xi - x, \lambda)\, d\xi$$

converged in the mean to $f(x)$. "It will be noticed that $f_\lambda(x)$ may be regarded as consisting of all the components of $f(x)$ with period not exceeding $2\pi\lambda$, together with a portion of the components of $f(x)$ with period exceeding $2\pi\lambda$ but not $2\pi(\lambda + 1)$." [Wiener 1925b.]
[11] Bochner remarked [1932, p. 114]: "Konsequenter aber typographisch umständlicher wäre die Schreibweise

$$\int e^{ixa} \frac{d^k E(\alpha, k)}{d\alpha^{k-1}} \quad \text{statt} \quad \int e^{ixa}\, d^k E(\alpha)."$$

[12] Cf. Ch. 2, §48 and note 67.
[13] In Ch. 2 we found such smoothing processes employed by the following authors: Evans [1928], Leray [1934], Friedrichs [1939], Calkin [1940] and Bochner [1946]. (See especially Ch. 2, note 66.)
[14] The operation of differentiation was indirectly present in the notation, but it was not defined explicitly. In [§30.4] a convergence argument is used. The convergence of $\varphi_n(\alpha)$ in F_k was used to conclude something about $d^k\varphi(\alpha)$ where $\varphi(\alpha)$ is the limit of the $\varphi_n(\alpha)$s. However, the convergence was not explicitly transferred to the symbols $d^k\varphi(\alpha)$.
[15] (43) and (44) only converge for β in formula (42) less than one. For $\beta \geq 1$ a slightly different definition was given:

$$G^{(m)}(z) = \frac{(-i)^m}{\sqrt{2\pi}} \int_L e^{-izy} y^m f_1(y)\, dt,$$

$$H^{(m)}(z) = \frac{(-i)^m}{\sqrt{2\pi}} \int_{L'} e^{-izy} y^m f_2(y)\, dy,$$

which determines G and H modulo a polynomial of degree $(m - 1)$.
[16] If the pair f_1, f_2 represents a tempered function, i.e. a function, satisfying (39) as explained in (41), then the Fourier transformed pair can be found from (38).
[17] As will be shown below the two theories are not isomorphic. Carleman's theory is the more general of the two.
[18] It is remarkable that the theory of hyperfunctions did not emerge from Carleman's work but in connection with Fantappié's theory of analytic functionals (Ch. 1, §6–8).

In Fantappié's theory of analytic functionals, the fundamental theorem was the representability of an analytic functional by a function (the indicatrix) defined and analytic in the complement of the domain of the functional (Ch. 1, §6). This theorem was developed further by G. Köthe [1951, 1952], among others (see also references there).

For an open subset of the Riemann sphere Ω Köthe defined the space $P(G)$ to be the space of functions analytic in the domain G and vanishing at ∞ if ∞ belongs to G.

In $P(G)$ he introduced a Fréchet topology in the sense of Dieudonné and Schwartz [1949]. He also considered the spaces $R(A)$ of functions f defined and analytic in an open set $D(f)$ containing the closed subset A of Ω and vanishing in ∞ if $\infty \in D(f)$. On the space $R(A)$ he defined a locally convex topology in a way similar to Schwartz' definition of the LF topology on $\mathscr{D}$. Köthe's main theorem then states that the dual $R'(A)$ of $R(A)$ is topologically isomorphic with $P(\Omega \backslash A)$ and conversely the dual $P'(Q - A)$ of $P(\Omega \backslash A)$ is isomorphic with $R(A)$. This theorem was independently derived by Grothendiek [1953] and Dias (cf. Köthe [1951, p. 30]).

The isomorphism used in both cases takes the functional F into its indicatrix

$$f(\lambda) = F\left(\frac{1}{z - \lambda}\right) \tag{1*}$$

from which F can inversely be expressed

$$F(g) = \oint f(\lambda) g(\lambda)\, d\lambda \tag{2*}$$

integrated over a suitable curve.

Köthe's and Fantappié's theories differ in that Köthe has been able to sharpen Fantappié's theory by applying Schwartz' and Dieudonné's recent abstract theory of duality of Fréchet spaces [1949]. This application suggested to Köthe the analogy between Fantappié's theory and the theory of distributions. Dieudonné mentioned to him that the connection was in fact that of an inclusion.

Köthe had already remarked in [1951] that

> Die hier entwickelte Theorie steht in engem Zusammenhang mit der Theorie der Distributionen von L. Schwartz, sie kann sogar, vorauf mich Herr Dieudonné aufmerksam machte, durch Einbettung des Raumes $P(G)$ in den Raum der im reellen Sinn unendlich oft differentierbaren komplexen Funktionen auf G aus ihr abgeleitet werden. [Köthe 1951.]

Thus distributions with compact support were special cases of analytical functionals.

In [1952] Köthe developed this idea further. This time he chose the closed space A to be a sufficiently nice curve C which did not pass through ∞.

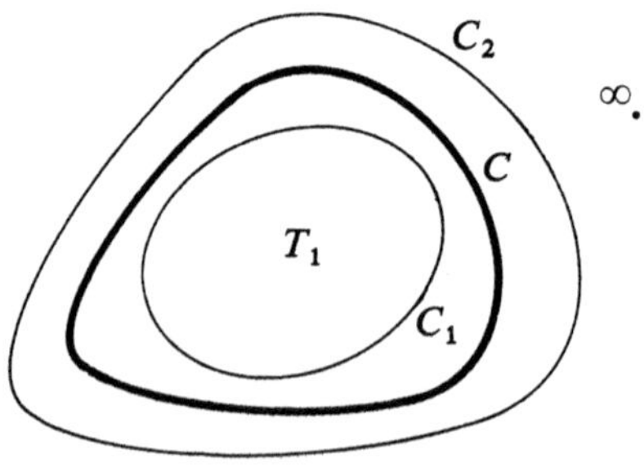

In this case the indicatrix of a functional in $R'(C)$ is described by two analytic functions f_1 and f_2 in $P(T_1)$ and $P(T_2)$, respectively, where T_1 and T_2 describe the two domains into which C separates the Riemann sphere. Thus analytic functionals on C are isomorphic with pairs of analytic functions.

Köthe showed that the analytic functional in a certain sense was a "Randverteilung", i.e. was a sum of the boundary values of the two indicatrices f_1 and f_2. Thus, if the functional F was given from the analytic function f by the integral

$$F(g) = \oint_C f(z)g(z)\,dz \tag{3*}$$

(cf. the way L^1 functions define distributions), then f_1 and f_2 have analytic continuations $\bar{f}_1, \bar{f}_2$ to C and $f = \bar{f}_1 + \bar{f}_2$ on C.

More generally a functional in $R'(C)$ is according to (2*) given by

$$F(g) = \int_{C_1} f_1(z)g(z)\,dz + \oint_{C_2} f_2(z)g(z)\,dz,$$

where C_1 and C_2 are curves in T_1 and T_2, respectively. Therefore one can still think of F as a sum of the "boundary values" of f_1 and f_2 (a more subtle convergence theorem is proved by Köthe).

In particular the distributions on C are analytic functionals and therefore Köthe concluded:

Randverteilungen auf C stellen also eine Verallgemeinerung des Distributionsbegriffs dar.

Köthe's [1952] was a great step in the theory of hyperfunctions. However his theory had the great disadvantage that it only generalized distributions on curves which did not contain ∞. Thus the distributions on the real line were excluded.

Inspired by Köthe, H. G. Tillmann extended the theory to include the real axis as well. Tillmann first proved that distributions with compact support were boundary values (jumps between two boundary values) of functions analytic in the upper and lower half-planes [Tillmann 1953, p. 76] (see also Tillmann [1957]). Since he used the functions f_1 and $(-f_2)$, f_1 and f_2 being the indicatrices, he arrived at the *jump* $f_1 - (-f_2)$ instead of Köthe's sum $f_1 + f_2$. In [1961a, p. 13] he extended the result to $\mathscr{D}'_{L^p}$ (i.e. derivatives of L^p functions) and in [1961b] to the tempered distributions, to the distributions of finite order, and to all of $\mathscr{D}'$. He characterized the growth conditions which the analytic functions must satisfy in order to determine distributions from those different classes of distributions. Independently M. Sato had developed the same theory [Sato 1958]. He generalized the theory of function pairs to higher dimensions in his famous article, "Theory of hyperfunctions" [1959/60], after A. Weil had informed him about Köthe's work. He called his generalized functions: hyperfunctions.

The theory of hyperfunctions has already proved valuable in physics [*Hyperfunctions and Theoretical Physics*, 1975] and is still developing. Its main power compared with the theory of distributions is that the strong results in complex analysis can be applied. For a treatment of hyperfunctions see, for instance, Schapira [1970].

In the early treatments of hyperfunctions by Köthe, Tillmann and Sato, the connection with Carleman's theory was neglected. It was pointed out by Bremermann and Durand [1961].

[19] For a complex number $z = x + iy$ in the lower half plane $(y < 0)$

$$|f(z)| = |f_2(z)| = \exp\,[(\log\sqrt{x^2+y^2})^2]\cdot\exp\,[-\arg\,(x+iy)]\cdot\exp\,(-y)$$
$$\times\exp\left[\frac{x^2+y(y+1)}{x^2+(y+1)^2}\right]$$

For fixed $y \neq 0$ the last three factors remains bounded away from 0 whereas the first factor tends to infinitely as quickly as $e^{(\log x)^2}$. Therefore the condition (48) cannot be satisfied, i.e., f does not represent a tempered distribution. For $z = re^{i\theta}(\sin\theta = K)$ the norm of f is

$$\exp\left[-Kr + \left(\log^2 r - \theta^2\right)\frac{r^2 + Kr}{r^2 + 2Kr + 1} + 2\theta \log r \frac{\sqrt{1 + K^2}r}{r^2 + 2Kr + 1}\right],$$

which for $\theta \neq 0$ satisfies Carleman's conditions. (I thank Professor Duistermaat (Utrecht) for this example.)

[20] I do not know whether Carleman's function pairs under the conditions (42) always represent distributions. Tillmann's growth conditions in [1961b] suggest that this is not the case.

I have not been able to rigorously prove that Carleman's and Schwartz' Fourier transforms of a tempered distribution are equal; but formal calculations strongly suggest that this is the case.

Chapter 4

[1] Courant–Hilbert's Green's function $K(x, x')$ has in modern terminology the property that

$$L_x(K(x, x')) = -\delta(x - x').$$

Thus it differs from the ordinary Green's function by a factor (-1). In what follows Courant's notation has been changed so as to correspond to the usual sign convention.

[2] Duhamel gave a similar argument in [1847a, b] where he expressed a solution to a partial differential equation (the heat equation) with variable (in time) boundary values as a superposition of solutions with constant boundary values (see Lützen 1979 III, 4). This is the famous Duhamel's principle.

[3] The fact that the theory of distributions transforms (if handled with care) a physical analysis into a rigorous mathematical proof is no doubt one of its great powers. In this respect the introduction of distributions parallels the creation of the calculus. The latter—in its Weierstrassian version—gave a rigorous method for transforming the physical analytic method of infinitely small quantities, as given for example in Archimedes' method (unknown to the inventors of the calculus) into a rigorous proof. Thereby the application of a separate synthesis—the exhaustion method—could be avoided. It seems to be of great advantage to a mathematical theory that it is close to physical intuition.

[4] This does not become clear in Green's presentation which makes the formulas slightly difficult to understand for a modern reader thinking in distribution terms. In loose distribution language (12) is still valid for the Green function U, but

$$\int_\Omega V\Delta\frac{1}{|\bar{x} - \bar{x}'|}\,d\bar{x}$$

becomes

$$-\int V 4\pi\delta(\bar{x} - \bar{x}')\,d\bar{x} = -4\pi V(x'),$$

giving the extra term on the right-hand side. The fact that only one formula (12) is needed for both cases gives the treatment in the theory of distributions a certain simplicity compared with the classical theory. However the simplicity should not be overestimated. The technicalities remain the same; they are only moved from one place in the proof to another. Thus the proof of (14) involves exactly the ball-cutting procedure used for the proof of (15) [Schwartz 1950/51, p. 45, Ex. 2].

5 It is interesting that although Kirchhoff's work on the Huygens principle was highly admired and cited by his successors, nobody explicitly refers to the particular form of F. Volterra and Hadamard abandoned Kirchhoff's procedure for other reasons than his use of this paradoxical function.

6 In the nonanalytic case Hadamard showed that the two equations differed profoundly in that they did not have the same type of well-posed problems [Hadamard 1932, Livre I, Ch. II].

7 Hadamard introduced the term "solution élémentaire" instead of fundamental solution. This usage is still maintained in French mathematical literature.

8 Zeilon also struggled with other types of singularities that make the integrals involved singular. He handled these difficulties by introducing complex integration in a way similar to the one used by Hadamard in one of his definitions of the partie finie (note 9).

9 Hadamard also defined the partie finie as a complex integral [Hadamard 1932, §80 and §82].

10 A less *ad hoc* definition of an integral similar to Hadamard's was given by Marcel Riesz in his penetrating studies of the Riemann–Liouville integral during the period 1933–1936. His results were published first in [1938/40] and later in a more comprehensive form in [1949]. He multiplied Hadamard's integral by the factor $1/[\Gamma(p+1)]$ and observed that the resulting integral (Riemann–Liouville's integral)

$$I^\alpha f(x) = \frac{1}{\Gamma(\alpha)} \int_a^x f(t)(x-t)^{\alpha-1}\, dt,$$

which is convergent for $\alpha > 0$, had an analytic extension to the negative real axis $\alpha \leq 0$. Riesz showed that the operator I^α satisfied

$$I^\alpha I^\beta = I^{\alpha+\beta} \quad \text{and} \quad \frac{d}{dx}(I^{\alpha+1}) = I^\alpha. \tag{$*$}$$

Schwartz [1950/51, Vol. I, p. 50 and Vol. II, p. 32] also studied the Riemann–Liouville integral and similar integrals, but he did not use them as an operator but as a functional, just as he did with the partie finie. Also for Schwartz the relations (∗) remained important.

11 This is only true for the period after the implementation of rigor between 1830 and 1870 approximately. Before that time arguments based on the δ-function were considered valid—see the arguments used in the proofs for the "convergence" of the Fourier series.

12 The idea of point charges and masses as mathematical idealizations makes the above-mentioned procedure in the treatment of electrical and gravitational forces even more logically questionable.

13 Maxwell's argument is based on the assumption that the only forces working on an atomic scale are the electrostatic forces. The discovery of the strong and weak interactions therefore invalidates the argument.

14 The first example is taken from the place in Weber and Gans' statistical mechanics

[1916] in which they showed how to introduce a temperature variable on a statistical basis [§257]. For a physical system they defined the function:

$$V(\varepsilon_*) = \int_{\varepsilon<\varepsilon_*} dx_1\, dx_2 \cdots dx_n, \tag{1*}$$

i.e. the volume of the points in phase space having an energy $\varepsilon < \varepsilon_*$. They wanted to show that the function

$$t = \frac{V(\varepsilon_*)}{\omega(\varepsilon_*)}, \tag{2*}$$

where

$$\omega(\varepsilon_*) = \frac{dV(\varepsilon_*)}{d\varepsilon_*} \tag{3*}$$

gave a sensible temperature measure. A property which a temperature variable φ must necessarily have is the following:

> (Vereinigungssatz) Bei Vereinigung zweier Systeme mit den gleichen Werten φ entsteht ein neues System mit eben demselben Werte φ. D.h.: Bezeichnet Σ den aus σ_1 und σ_2 zusammengesetzten Mechanismus, so muss, falls
>
> $$\varphi_{\sigma_1}(\varepsilon_1) = \varphi_{\sigma_2}(\varepsilon_2) \tag{4*}$$
>
> ist, auch
>
> $$\varphi_\Sigma(\varepsilon_1 + \varepsilon_2) = \varphi_{\sigma_1}(\varepsilon_1) = \varphi_{\sigma_2}(\varepsilon_2) \tag{5*}$$
>
> sein.

In order to prove this theorem Weber and Gans introduced the product

$$f(\varepsilon_2) = V_1(E - \varepsilon_2)V_2(\varepsilon_2) \tag{6*}$$

from which they defined the expressions

$$V_{12}(E) = \int_0^E f(\varepsilon_2)\frac{t_1(E-\varepsilon_2) + t_2(\varepsilon_2)}{2t_1(E-\varepsilon_2)t_2(\varepsilon_2)}\, d\varepsilon_2, \tag{7*}$$

$$\omega_{12}(E) = \int_0^E f(\varepsilon_2)\frac{1}{t_1(E-\varepsilon_2)t_2(\varepsilon_2)}\, d\varepsilon_2, \tag{8*}$$

where the indices 1, 2 and 12 refer to the systems σ_1, σ_2 and Σ, respectively, and $E = \sigma_1 + \sigma_2$ is the energy of the compound system Σ.

> Aus den beiden letzten Gleichungen erhält man $t_{12}(E)$ durch Division (2*). Nun würde sich diese ohne weiteres ausführen lassen, wenn es erlaubt wäre, die ungleichen Faktoren der Integranden vor das Integral zu ziehen; dies Verfahren wird aber auch dann zu einem angenähert richtigen Ergebnis führen, wenn ihre gleichen Faktoren für ein Argument einen überwiegend grossen Wert besitzen.

They can argue that f has in fact one maximum $\hat{\varepsilon}_2$ "und die Funktion f sei an dieser Stelle ausserordentlich steil" (Figure 1*).

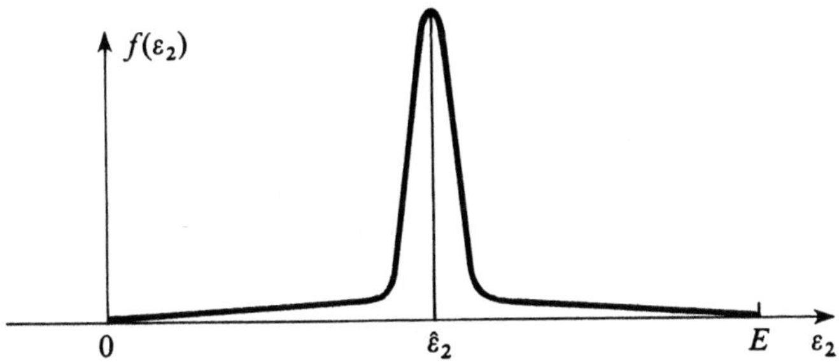

Figure 1*

Therefore "können wir in (7*) und (8*) in dem t enthaltenden Faktor ε_2 durch $\hat{\varepsilon}_2$ ersetzen und bekommen:

$$t_{12} = \frac{V_{12}(E)}{\omega_{12}(E)} = \tfrac{1}{2}[t_1(E - \hat{\varepsilon}_2) + t_2(\hat{\varepsilon}_2)]\text{''}. \tag{9*}$$

Differentiation of f in (6*) shows that the maximum value $\hat{\varepsilon}_2$ satisfies

$$t_1(E - \hat{\varepsilon}_2) = t_2(\hat{\varepsilon}_2). \tag{10*}$$

If now $t_1(\varepsilon_1) = t_2(\varepsilon_2)$, one sees that $\hat{\varepsilon}_2 = \varepsilon_2$ satisfied (10*). Hence from (9*)

$$t_{12} = \tfrac{1}{2}[t_1(\varepsilon_1) + t_2(\varepsilon_2)] = t_1(\varepsilon_1) = t_2(\varepsilon_2),$$

so that the "Vereinigungssatz" is proved.

As a second example we choose a section in J. Bernamont's "Fluctuations de potentiel aux bornes d'un conducteur métallique de faible volume parcouru par un courant" [1937], in which he treated spectral theory of fluctuating quantities. In one place in his calculations he was faced with the problem of finding the integral

$$\int_0^\infty f''(t) \cos 2\pi\nu t \, dt, \tag{11*}$$

where $f(t)$—the so-called correlation function—was known to be

$$f(t) = Ae^{\alpha|t|} \quad \text{for } |t| > t_0 > 0, \tag{12*}$$

but not known in the small interval $[-t_0, t_0]$. He wrote

$$f(t) = f_1(t) + \varphi(t), \tag{13*}$$

where $f_1 = Ae^{\alpha|t|}$ and $\varphi = f - f_1$ is a quickly varying function around $t = 0$, zero outside $[-t_0, t_0]$ but not known in greater detail. f was supposed to be C^∞. (Figures 2*, 3* and 4* are copied from Bernamont [1937].)

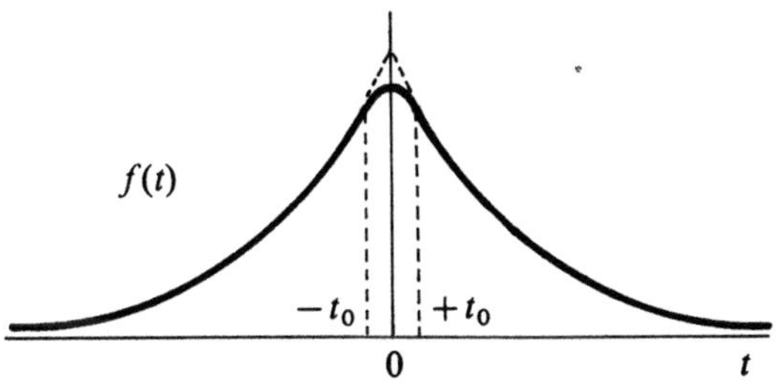

Figure 2*

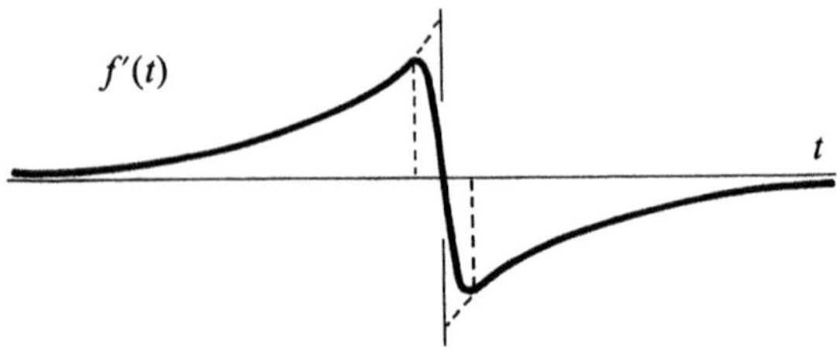

Figure 3*

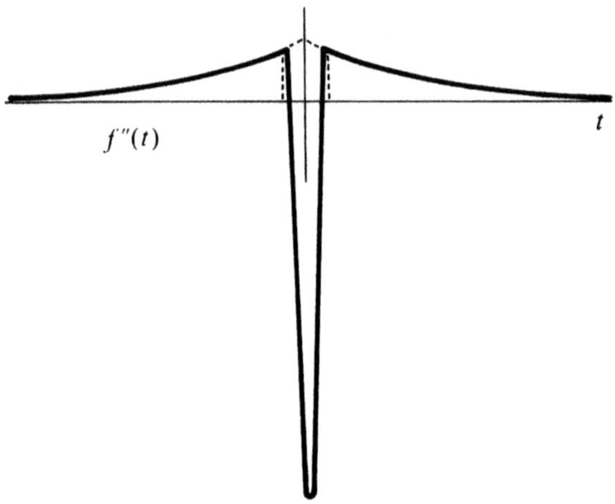

Figure 4*

Thus the integral (11*) became

$$\int_0^\infty \varphi''(t) \cos 2\pi\nu t \, dt + \int_0^\infty (Ae^{\alpha t})'' \cos 2\pi\nu t \, dt. \tag{14*}$$

The second integral in (14*) is a known quantity. "Le premier terme s'intègre seulement entre 0 et t_0 intervalle, ou le cosinus peut être considéré comme constant et égal à 1." Thus

$$\int_0^\infty \varphi''(t) \cos 2\pi\nu t \, dt = -\varphi'(0) = f'_1(0), \tag{15*}$$

the last equality being found from (13*) when $f'(0) = 0$. This calculation corresponds to the fact that in the distribution sense:

$$f''_1 = [f''_1] + 2f'_1(0+)\delta, \tag{16*}$$

where $[f''_1]$ is the *function* equal to f''_1 for $t \neq 0$ and equal to 0 at $t = 0$. (In (15*) the factor 2 is absent because Bernamont only integrated from 0 to ∞.) However Bernamount deliberately chose the approximation f, having only what he called a pseudo-discontinuity of the second order, instead of the more obvious approximation f_1, having a real discontinuity of the second order. The reason for this no doubt was that it would be very difficult for him to give an argument for the additional term $f_1'(0)$ when the integral is only taken over $[0, \infty]$.

In the step from (11*) to (14*) Bernamount leaves out without comment such a contribution from the peak of f_1 at 0. The result is correct because $\varphi(t)$ has a similar but inverted peak in 0, a fact immediately seen from (13*).

These two examples from the beginning of this century illustrate that the δ-function was so urgently necessary for mathematical physics that it presented itself in disguise also in cases where a special δ-function was not "defined".

[15] In this note I want to correct a misunderstanding concerning the early use of the δ-function which may arise from a note in Youschkevich's paper, "The concept of function up to the middle of the 19th Century" [1976] Concerning Euler's last memoir on aerial motion [Euler 1765c], Youschkevich writes [p. 71]:

> To study the solutions of the functional equation [governing the motion] ... he [Euler] introduces functions that have the value 0 at all points except one. He remarks that since these pulse functions form what is called now a (non-enumerable) basis for the set of all functions, use of them as initial values for a wave function makes it possible to describe concisely and in geometric terms the entire theory of propagation and reflection of plane waves.

The reader of this note is left with the impression that Euler used δ-functions and knew the formula

$$(f * \delta)(x) = \int f(a)\delta(x - a)\, da = f(x). \tag{1*}$$

This, however, is far from the truth. Youschkevich's reference is to the place in which Euler showed how an initial distortion, in a limited part of a thin infinitely long organ pipe propagates in time. The (infinitely small) velocities v in the positive direction and the density q are in their initial states, represented by the curves InK and ImK, respectively,

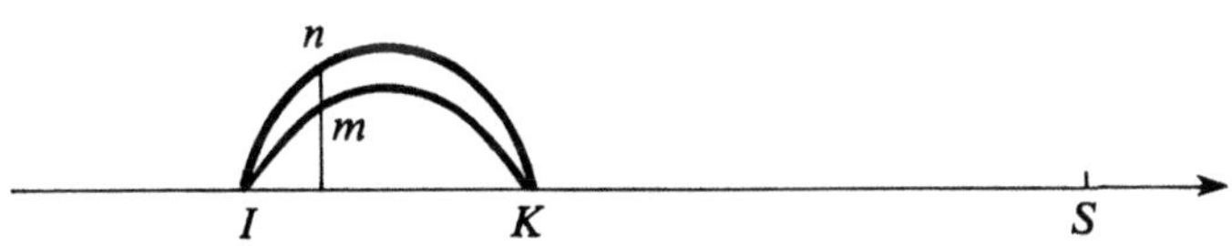

Figure 1*

in the sense that at the point t: $v = tn$ and $q = b + (b/c)(tm)$, where b is the "natural" density of air and c is the speed of propagation. Euler proved that at a time $\tilde{t}$ the values of v and q in S are determined by

$$\begin{aligned} v &= \tfrac{1}{2}(tn \pm tm) && \text{where } tS = c\tilde{t}, \\ q &= b + \frac{b}{2c}(tm \pm tn) && (+ \text{ used if } S \text{ is to the right of } t \\ & && - \text{ used if } S \text{ is to the left of } t). \end{aligned} \tag{2*}$$

Euler showed how these formulas together with the method of images could account for the motion, also in the case of a semifinite pipe. In the case of a pipe bounded at both ends however Euler found Figure 1* too complicated:

Ensuite, pour ne pas trop embrouiller les idées, je conçois l'agitation initiale comme faite dans un seul point t, la densité y étant $= b + (b/c) \cdot tm$ et la vitesse $= tn$, et que partout ailleurs sur AB les deux appliquées évanouissent.

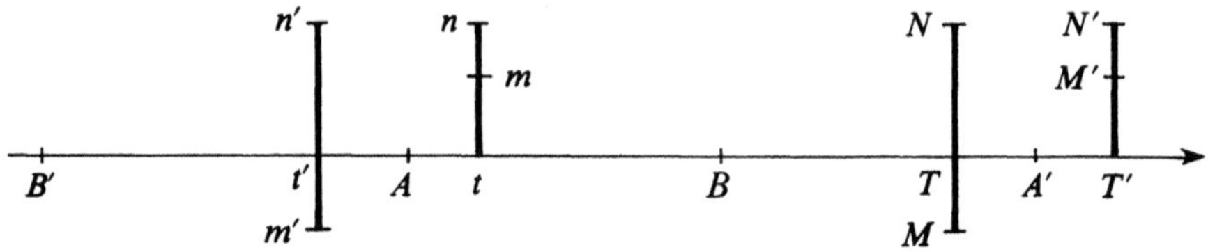

Figure 2*

Again the images as shown in Figure 2* together with (2*) determines the motion.

Thus Euler's pulse functions were not δ-functions but functions with support at one point at which their values were finite. Euler has probably observed that in (12*) only the value of the function at two points was used; therefore clarity could be gained by isolating these values and neglecting the rest. It might be argued that Euler saw that if the pulse functions were drawn for all t in IK, then the original curves would arise (in modern terms this corresponds to the pulse functions being a basis). However Euler did not say so explicitly in [1765c].

Youschkevich's information on Euler's use of pulse functions comes from Truesdell whose introduction to Euler's *Opera Omnia*, ser. II, Vol. 13, p. LXII [1960] he cites. Truesdell showed that the pulse functions formed a basis for the set of functions, without attributing the argument to Euler. Concerning Euler's argument he wrote:

> It is the first occurrence of the type of argument since become familiar in connection with the "delta functions". There can be no logical objections here, however: for the simple wave theory, the argument is rigorous. That is, if we put
>
> $$f_\alpha(x) = \begin{cases} f(x), & \text{when } x = \alpha, \\ 0, & \text{when } x \neq \alpha, \end{cases} \tag{3*}$$
>
> then for any given interval we have
>
> $$f(x) = \sum_\alpha f_\alpha(x), \tag{4*}$$
>
> where α runs over the interval.

The reason why such f_αs may be used here instead of δ-functions is that only the point evaluation in (2*) are made; if integration had been involved, the splitting (3*) would not have worked since $f_\alpha = 0$ in L^1.

I conclude that contrary to what Youschkevich's note seems to indicate, Euler did not use δ-impulses or the formula (1*) in the referred argument.

16 The square bracket is another expression of Dirichlet's kernel:

$$\frac{\sin\,(i + \frac{1}{2})(\alpha - x)}{2 \sin \frac{1}{2}(\alpha - x)}.$$

17 Pringsheim [1907] pointed out that the correct Fourier integral theorem:

$$f(x) = \frac{1}{\pi} \int_0^\infty \int_{-\infty}^\infty f(\alpha) \cos p(\alpha - x)\, d\alpha\, dp$$

cannot be found in Fourier. Pringsheim credits Cauchy [1827] with the first appearance of the correct formulation. Pringsheim called Fourier's version (87) "sinnlos" and a "fehlerhafte Schreibweise".

[18] Before Fourier, Lagrange had offered a similar argument in his "Recherches sur la nature et la propagation du son" [1759, §38]. In this article Lagrange came formally close to the Fourier series. His form of the Fourier series looks like (81), but is complicated by the fact that $d\alpha$ is replaced by another infinitesimal which depends on the number of terms in the series. In this way the integration, which is not a real integration, and the summation are made dependent on one another. After strange manipulations with infinite quantities he "shows" that a certain kernel has the δ-function property described by Fourier in (82). The result corresponds to the formula:

$$\frac{\sin\dfrac{\pi X}{2a}\sin\dfrac{\pi}{2}\left(\dfrac{nx}{a}+\dfrac{nHt}{T}\right)}{\cos\dfrac{\pi}{2}\left(\dfrac{x}{a}+\dfrac{Ht}{T}\right)-\cos\dfrac{\pi X}{2a}} \to a\delta(x+ct-X).$$

Euler [1759] wrote that Lagrange in this article had used "des calculs qui paroissent tout a fait indechiffrable". It would be an interesting but not easy task to give an explanation of what precisely is happening in Lagrange's calculation, which is indeed strange.

Another Fourier expansion of an impulsive function can be found in Lagrange's discussion of the propagation of sound in air [1759, §54]. Both places in Lagrange have been studied by Ravetz [1961].

[19] See, for example, references in Bremmer and van der Pol [1955, pp. 62, 65] to Hermite [1891] [cited there] and Lebesgue.

Lebesgue [1906, p. 74] defined singular integrals as follows:

> Il existe toute une classe de fonctions $\varphi(n, \theta)$ des deux variables n et θ qui jouissent des propriétés suivants: 1° pour une certaine valeur x de θ la fonction $\varphi(n, \theta)$ croit indéfiniment avec n; 2° $\int_\beta \varphi(n, \theta)\, d\theta$ tend vers zéro avec n pourvu que x n'appartienne pas à l'intervalle (β, γ); 3° au contraire cette intégrale tend vers une limite déterminée et non nulle quand n croit, si x est intérieure à l'intervalle (β, γ); cette limite est alors évidemment indépendante de β et γ.

[20] Inspired by Heaviside, W. E. Sumpner [1931] gave a trigonometric definition of $H(t)$ and remarked:

> It thus appears that the $H(t)$ function is based throughout upon Fourier's theorem, and that, when the latter is expressed in impulsive form [this is Sumpner's expression for the formula $f(x) = \int_{-\infty}^{\infty} f(t)(d/dt)H(t-x)\,dt$], what is done is not to establish a new theorem, but to vary the statement of an old one. [Sumpner 1931, p. 363.]

Sumpner tried by an infinitesimal method to rescue Heaviside's treatment of Fourier series and δ-functions. This attempt will be discussed in §43 and §44.

[21] Some of the formulas in the tables of Fourier transforms involving the δ-function were already contained in Fourier's and Heaviside's earlier works discussed in §18, §19, and §22 (especially (109)).

[22] Dirac's notation changed from one edition of his book to another. In the first edition the vectors ψ are called ψ-symbols and the operators are just called observables. I shall use modern terminology but otherwise remain faithful to the first edition.

[23] See Jammer [1966] for the role of Dirac's book in the conceptual development of quantum mechanics.

[24] In this he differed from Banach, who called his spaces B-spaces, and left it to posterity to add "anach".

[25] In Wenzel's book on quantum field theory [1943] the δ- and the Δ-functions are used as powerful tools [see pp. 20–26, in particular]. Dirac made use of the distribution (156) in his explanation of the positron as a hole in an otherwise full "sea" of electrons in states with negative energy. In Dirac's work [1934] which preceded Pauli's use of (156) one also finds (147). [Dirac 1933] is not his first paper on the positron; see[Hanson 1963].

[26] In [Courant and Hilbert 1937] pp. 443–448 the solution of the equation (155) with the boundary conditions

$$u(\bar{r}, 0) = 0 \quad \text{and} \quad \frac{d}{dt} u(\bar{r}, 0) = \psi$$

is shown to be

$$u(\bar{r}, t) = \frac{1}{4\pi t} \frac{\partial}{\partial t} \iiint_B J_0(\sqrt{t^2(\bar{r}' - \bar{r})^2})\psi(\bar{r}')\, d\bar{r}',$$

where B is that part of the (x, y, z) hyperplane which lies inside the light cone with vertex in $(\bar{r}, t)$.

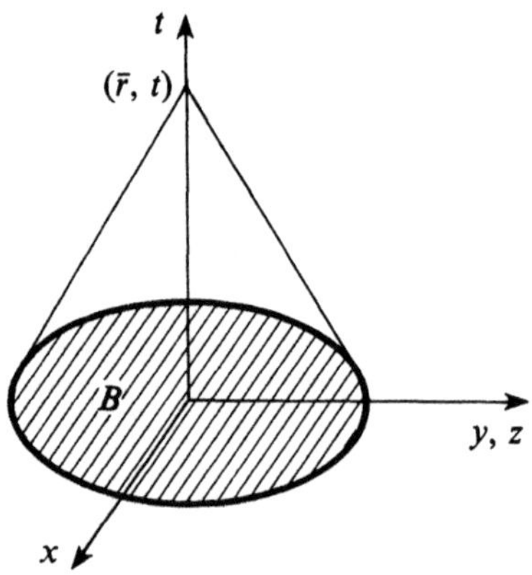

This corresponds to a Green's function, i.e. a solution to (155)

$$u(\bar{r}, t) = \frac{1}{4\pi t} \frac{\partial}{\partial t} [J_0(\kappa\sqrt{t^2 - r^2}) H(t^2 - r^2)]$$

which is the same as (156) with

$$F(\bar{r}, t) = \begin{cases} J_0(\kappa(t^2 - r^2)^{1/2}) & \text{for } t > r \\ 0 & \text{for } r > t > -r \\ J_0(\kappa(t^2 - r^2)^{1/2}) & \text{for } -r > t. \end{cases}$$

Courant–Hilbert's solution thus has the wrong sign for $t < 0$, but for $t > 0$ it corresponds to Pauli's formulas (156) and (157). An examination of Courant–Hilbert's proof shows that they implicitly assume t to be positive.

[27] If $g_n \rightrightarrows f$ uniformly in $[a, b]$ then the k-times iterated integrals

$$\int_a \int_a \int_a \int_a \cdots g_n(x)\,dx \rightrightarrows \int_a \int_a \int_a \int_a \cdots f(x)\,dx \quad \text{uniformly in } [a, b].$$

Given $f \in \mathscr{E}$ and k. According to Weierstrass' approximation theorem, there exists a sequence ${}_ip_k$ of polynomials such that

$${}_ip_k \rightrightarrows f^{(k)} \text{ in } [-k, k] \quad \text{for } i \to \infty.$$

Thus, according to what is above, the k-times integrated functions

$$\iiint {}_ip_k = {}_iq_k$$

satisfy

$${}_iq_k^{(j)} \rightrightarrows f^{(j)} \text{ on } [-k, k] \quad \text{for } 0 \le j \le k.$$

Choose i so large that ${}_iq_k = q_k$ satisfy

$$|q_k^{(j)}(x) - f^{(j)}(x)| < \frac{1}{k} \quad \text{for } x \in [-k, k].$$

Then the sequence of polynomials q_k will converge to f in $\mathscr{E}$.

[28] Evans, who was the first to use general measures or additive functions of points in potential theory, wrote in the introduction to his article:

> The Stieltjes integral [an integral with respect to an arbitrary measure] is well adapted to the investigation of problems in mathematical physics first because it applies equally well to discrete and continuous sequences of values, and thus enables either to be regarded as an approximation to the other, and in the second place because it is based on additive functions of point sets, or in special cases additive functions of points, curves and surfaces, of limited variation. These latter are familiar to us in volume, point curvelinear and surface distributions of mass and electricity.

As discussed in Ch. 2, §33, Evans was able to describe such line and point distributions in two dimensions with the help of the general measures and Stieltjes integrals. For example, he could show that for a point mass in M'', corresponding to the additive function of point sets:

$$f(e) = \begin{cases} 1 & \text{for } m \in e, \\ 0 & \text{for } m \notin e, \end{cases}$$

the potential

$$\log \frac{1}{MM''} = u(M) = \frac{1}{2\pi} \int_\Sigma \frac{1}{\log MM'}\, df(e')$$

was a solution to the generalized Poisson equation

$$\int_S \nabla_n u\, ds = f(s),$$

where $f(s)$ is described in Ch. 2, note 48.

[29] Indeed it does if a distribution is defined as a pair consisting of a differential operator and a function, as Tolhoek and others later defined it (see Appendix, §3).

[30] Smith referred to a discussion in *Nature*, Vols. 58–60 between Michelson, Gibbs and Love among others on summation of Fourier series and the Gibbs' phenomenon. In this interesting discussion, which took place in the column "Letters to the Editor", it becomes clear that some physicists represented by Michelson thought it inadequate that a Fourier series of a function at a point x of discontinuity converged towards the mean value of $f(x + 0)$ and $f(x - 0)$ and not towards all the points on the vertical line segment joining $(x, f(x + 0))$ and $(x, f(x - 0))$, which is obviously a part of the limiting curve for the graphs of the partial sums in the Fourier series.
[31] Multiplication of an improper function $D_1(x_1, \ldots, x_n)$ and another $D_2(x_1, \ldots, x_m)$ was defined as an improper function in $(m + n)$ variables.

Chapter 5

[1] I have chosen to treat this theory of currents in a separate chapter not because it is of particular importance but because it does not fit into any of the other chapters.
[2] Notation.

A *chain element* of dimension p in a closed orientable differentiable manifold V of dimension n is the image of a polygon π in $\mathbb{R}^p$ under a C^r function μ, where $\mathbb{R}^p$ or μ is oriented (the orientation, which is of great interest in the use of the theory, is not significant for the ideas in which we are interested, and will thus be neglected as much as possible).

A *chain* is a formal real linear combination of chain elements. The *boundary of a chain element* is equal to $\mu(b\Pi)$ where $b\Pi$ is the ordinary boundary of the polygon Π. It is a chain. The *boundary of a chain* is the proper linear combination of the boundaries of the chain elements in the chain.

A *p-form* or a *form of degree p* is an expression:

$$\omega = \sum_{i_1, \ldots, i_p} A_{i_1 \cdots i_p}\, dx_{i_1}\, dx_{i_2}\, dx_{i_3} \cdots dx_{i_p},$$

where the As are $C^1(V)$ functions of the coordinates, $x_1, \ldots, x_n$, on V.

The *integral* of a form ω over a chain element $C = \mu(\Pi)$ is defined as

$$\int_C \omega = \int_\Pi \mu * \omega \quad \text{(with a suitable orientation)},$$

where $\mu*$ acts on the coefficient functions f as follows:

$$(\mu * f)(y) = f(\mu y).$$

The integral of a form ω over a chain is defined by taking the linear combination of the integrals over the chain elements. A form ω is said to be *closed* (de Rham [1936] says exact) if

$$d\omega = 0.$$

A chain c is said to be *closed* and is called a *cycle* if

$$bc = 0.$$

[3] This is only true if we neglect orientation. Distributions and 0 currents are equal in the same sense as the functions f are equal to the forms $f\, dx_1, \ldots, dx_n$.

Chapter 6

[1] A Fréchet space E is called reflexive if the strong dual E_b'' of the strong dual E_b' of E is isomorphic with the space E itself via the mapping

$$x \rightarrow \text{value at } x.$$

[2] Two subsets of $\mathbb{R}^n$ are similar if they can be transformed into one another by a combination of a rotation, a dilation, and a translation. Two mass distributions (E_0, λ_0) and (E, λ) are called similar when E_0 is similar to E and similarly situated subsets of E_0 and E have the same masses.

[3] F is called polyharmonic if it is a solution to an equation

$$\Delta^n F = 0$$

for some n.

[4] All the information in this section stems from [Schwartz 1978, Interview]. This is the case with most of §7–§10 as well.

[5] In "Théorie générale des fonctions moyenne-périodiques" [Schwartz 1947a] the theory of distributions entered at some points. Differentiation in the distribution sense of functions is used throughout. Moreover, in §19 Schwartz briefly treated "distributions moyenne-périodiques". In this connection he defined the Fourier transform of distributions with compact support, but he did not define tempered distributions.

Concluding Remarks

> [1]... vi har haft forskellige særdeles interessante Besøg af udenlandske Matematikere, i første Linie den yngre franske Matematiker Prof. Laurent Schwartz, for hvis fremragende Bidrag til den klassiske Differential og Integralregning jeg—hvad der måske dog er unødvendigt—agter at propagandere stærkt over i Staterne.

[2] Already Hilbert saw this trend counteracting the diversification of modern mathematics. He expressed it as follows in his famous talk on "Mathematical problems" in 1900 [Hilbert 1900].

> Auch bemerken wir: je weiter eine mathematische Theorie ausgebildet wird, desto harmonischer und einheitlicher gestaltet sich ihr Aufbau, und ungeahnte Beziehungen zwischen bisher getrennten Wissenszweigen werden entdeckt. So kommt es dass mit der Ausdehnung der Mathematik ihr einheitlicher Charekter nicht verlorengeht, sondern desto deutlicher offenbar wird.

Also H. Weyl stressed this uniting force in the paper [1951] quoted above.

[3] Readers who are interested in the development of structural mathematics are referred to H. Mehrtens' excellent treatment of the history of lattice theory [Mehrtens 1979]. Lattice theory seems to be one of the less important structures. It is, according to Dieudonné [1978], an example of a generalization for the sake of generalization. "So much lattice and so few tomatoes," was Tom Lehrer's reaction to Birkhoff's lattice theory. However, the development of lattice theory is in many ways representative of mathematics in the twentieth century. In this book I have given an historical analysis of a more important but less typical theory.

[4] Even so the theory of differential equations and other areas in mathematical analysis have been less influenced by the structural movement than most of the other parts of mathematics. J. Fang [1970] in his description of the hierarchy of structural mathematics, writes:

> Farther along, at the lowest end of the structural totem pole, one finally descends upon the ground of the particular and individual where certain areas have long remained or will for some time remain indeterminate, structure-wise, ... For example, certain fragments from the theory of numbers, of functions of a real or complex variable, of differential equations, of differential geometry, etc.

[5] Marshall H. Stone in his article on "The Revolution in mathematics" [1961] is more extreme than Dieudonné in the emphasis placed on the purity and abstractness of modern mathematics. "Indeed, it is clear that mathematics may be likened to a game—or rather an infinite variety of games—in which the pieces and moves are intrinsically meaningless." Stone's article provoked Courant (see [Carrier 1962]) and others to warn against the separation of mathematics and science. That mathematics has now diverged from science more than ever before, has been denied by none. What Courant and others have argued against is the desirability of this state of affairs. See also the recent polemics by M. Kline [1973, Ch. 10; 1977].

Bibliography

d'Alembert, J.

[1747] Recherches sur la courbe que forme une corde tenduë mise en vibration. *Hist. et Mem. Acad. Sci. Berlin*, **3** (1747), pp. 214–249.

[1750] Addition au mémoire sur la courbe que forme un corde tenduë mise en vibration. *Hist. et Mem. Acad. Sci. Berlin*, **6** (1750), pp. 355–360.

[1761] Sur les vibrations des cordes sonores. *Opuscules Mathématiques*, **1** (1761), pp. 1–64 and Supplément, pp. 65–73.

[1780] Sur les fonctions discontinues. *Opuscules Mathématiques*, **8** (1780), pp. 302–308.

Anger, G.

[1961] Die Entwicklung der Potentialtheorie im Hinblick auf ihre grundlegenden Existenzsätze. *Jahresber. Deut. Math. Ver.*, **64** (1961), pp. 51–78.

Arbogast, L. F. A.

[1791] *Mémoire sur la Nature des Fonctions Arbitraires qui Entrent dans les Intégrales des Equations aux Différences Partielles*. St. Petersburg, 1791.

Banach, S.

[1932] *Théorie des Opérations Linéaires*. Varsovie-Lwow, 1932.

Belinfante, F. J.

[1946] On the longitudinal and the transversal delta-function, with some applications. *Physica*, **12** (1946), pp. 1–16.

Beltrami, E. J.

[1963] Some alternative approaches to distributions. *Siam. Rev.*, **5** (1963), pp. 351–357.

Berg, E. J.

[1929] *Heaviside's Operational Calculus*. New York, 1929.

Bernamont, J.

[1937] Fluctuations de potenti 1 aux bornes d'un conducteur métallique de faible volume parcouru par un courant. *Ann. Phys.*, Ser. 11, **7** (1937), pp. 71–140.

Bernkopf, M.

[1966] The development of function spaces with particular reference to their origins in integral equation theory. *Arch. Hist. Ex. Sci.*, **3** (1966), pp.1–96.

Beurling, A.

[1947] *Sur les Spectres des Fonctions*. Colloque Analyse Harmonique Nancy. CNRS. Paris, 1949, pp. 9–30.

Bôcher, M.

[1905/06] On harmonic functions in two dimensions. *Proc. Amer. Acad. Sci.*, **41** (1905/06), pp. 577–583.

Bochner, S.

[1927] Darstellung reelvariabler und analytischer Funktionen durch Verallgemeinerte Fourier- und Laplace-Integrale. *Math. Ann.*, **97** (1927), pp. 635–662.

[1932] *Vorlesungen über Fouriersche Integrale*. Leipzig, 1932.

[1946] Linear partial differential equations with constant coefficients. *Ann. of Math.*, **47** (2) (1946), pp. 202–212.

[1952] Review of L. Schwartz' *Théorie des Distributions*. *Bull. Amer. Math. Soc.*, **58** (1952), pp. 78–85.

Bochner, S. and Hardy, G. H.

[1926] Note on two theorems of N. Wiener. *J. London Math. Soc.*, **1** (1926), p. 240.

Bochner, S. and Martin, W. T.

[1948] *Several Complex Variables*. Princeton, 1948.

Bohr, H.

[1950] Address. *Proc. Intern. Congr. Math. 1950*, Vol. 1, p. 127.

Bourbaki, N.

[1948] L'architecture des mathématiques. In: *Les Grands Courants de la Pensée Mathématique*. Edited by F. le Lionnais (1948), pp. 35–47.

[1969] *Eléments d'Histoire des Mathématiques*. Paris, 1969.

Brelot, M.

[1972] Les étapes et les aspects multiples de la théorie du potential. *L'Enseign. Math.*, **18** (1972), pp. 1–36.

Bremermann, H. J.

[1965] *Distributions, Complex Variables and Fourier Transforms*. Massachusetts, 1965.

Bremermann, H. and Durand, L.

[1961] On analytic continuation, multiplication, and Fourier transformations of Schwartz' distributions. *J. Math. Phys.*, **2** (1961), pp. 240–258.

Bremmer, H. and van der Pol, B.

[1955] *Operational Calculus Based on the Two-sided Laplace-Integral*, 2nd ed. New York, 1955.

de Broglie, L.

[1948] Le role des mathématiques dans le dévelopment de la physique théorique contemporaine. In: *Les Grands Courants de la Pensée Mathématique*. Edited by F. le Lionnais (1948), pp. 398–412.

Browder, F. E.

[1975] The relation of functional analysis to concrete analysis in 20th century mathematics. *Hist. Math.*, **2** (1975), pp. 575–590.

Burkhardt, H.

[1914] *Trigonometrische Reihen und Integrale*. Encycl. Math. Wiss. II. A.12. Printed 1914.

Burkhardt, H. and Meyer, W. F.
[1900] *Potentialtheorie (Theorie der Laplace–Poisson'schen Differentialgleichung).* Encycl. Math. Wiss. 2. Bd. 1. Teil 1, pp. 464–503. Printed 1900.

Burkill, J. C.
[1926a] The Stieltjes integral in harmonic analysis. *Math. Gazette*, **13** (1926), pp. 195–196.
[1926b] The inversion formulae of Fourier and Hankel. *Proc. London Math. Soc.*, Ser. 2, **25** (1926), pp. 513–529.

Bush, V.
[1929] *Operational Circuit Analysis.* New York, 1929.

Calkin, J. W.
[1939] Abstract symmetric boundary conditions. *Trans. Amer. Math. Soc.*, **45** (1939), pp. 369–442.
[1940a] Functions of several variables and absolute continuity, I. *Duke Math. J.*, **6** (1940), pp. 170–186.
[1940b] Abstract definite boundary value problems. *Proc. Nat. Acad. Sci. USA* **26** (1940), pp. 708–712.
[1940c] Symmetric transformations in Hilbert space. *Duke Math. J.*, **7** (1940), pp. 504–508.
[1941] Two sided ideals and congruence in the ring of bounded operators in Hilbert space. *Ann. of Math.*, **42** (2) (1941), pp. 839–873.

Campbell, G. A.
[1928] The practical application of the Fourier integral. *Bell System Tech. J.*, **7** (1928), pp. 639–707.

Campbell, G. A. and Foster, R. M.
[1931] *Fourier Integrals for Practical Applications.* New York, 1931.

Cantor, G.
[1884] De la puissance des ensembles parfaits de points. Extrait d'une lettre adressée à l'éditeur. *Acta Math.*, **4** (1884), pp. 381–392.

Caratheodory, C.
[1904] Inaugural dissertation. Göttingen, 1904.
[1906] Über starke Maxima und Minima bei einfachen Integralen. *Math. Ann.*, **62** (1906), pp. 449–503.

Carleman, T.
[1944] *L'Intégrale de Fourier et Questions qui s'y Rattachent.* Uppsala, 1944.

Carrier, G. F., et al.
[1962] Applied mathematics: what is needed in research and education. *Siam Rev.*, **4** (1962), pp. 297–320.

Cartan, E.
[1928] Sur les nombres de Betti des espaces de groupes clos. *Comp. Rend. Acad. Sci. Paris*, **187** (1928), pp. 196–198.
[1929] Sur les invariants integraux de certains espaces homogènes clos et les proprietés topologiques de ces espaces. *Ann. Soc. Polon. Math.*, **8** (1929), pp. 181–225.

Casimir, H. B. G.
[1945] On Onsager's principle of microscopic reversibility. *Rev. Mod. Phys.*, **17** (1945), pp. 343–50, also *Phil. Res. Reports*, **1** (1946), p. 185.

Cauchy, A. L.

[1823a] *Resumé des Leçons Données à l'École Royale Polytechnique sur le Calcul Infinitésimal*, Tome 1. Paris, 1823 = *Oeuvres Complètes*. Paris, 1882–1958, **4** (2), pp. 5–261.

[1823b] L'intégration des équations linéaires aux différentielles partielles et à coefficients constants. *J. Ec. Polyt. Cah. 19*, **12** (1823), p. 511 = *Cauchy's Oeuvres Complètes*. Paris, 1882–1958, **1** (2), pp. 275–335.

[1827] Théorie de la propagation des ondes à la surface d'un fluide pesant, d'une profondeur indéfinie. *Mém. Acad. Roy. Sci. Inst. France, Sci. Math. et Phys.*, **1** (1827) = *Oeuvres Complètes*. Paris, 1882–1958, **1** (1), pp. 5–318.

[1841] Mémoire sur l'intégration des équations homogènes en termes finis. *Comp. Rend. Acad. Sci. Paris.* **13** (1841) = *Oeuvres Complètes*. Paris, 1882–1958, **6** (1), pp. 326–41.

Choquet, G. and Deny, J.

[1944] Sur quelques propriétés des moyennes, caractéristiques des fonctions harmoniques et polyharmoniques. *Bull. Soc. Math. France*, **72** (1944), pp. 118–141.

Christoffel, E.

[1876] Untersuchungen über die mit Fortbestehen linearer partieller Differentialgleichungen verträglichen Unstetigkeiten. *Anali di Mat.*, **8** (2) (1876) = Gesammelte Math. Abh. 2, pp. 51–80.

[1877] Ueber die Fortpflanzung von Stössen durch elastische feste Körper. *Anali di Mat.*, **8** (2) (1877), pp. 193–243 = Gesammelte Math. Abh. 2, pp. 81–126.

Courant, R. and Hilbert, D.

[1924] *Methoden der Mathematischen Physik, I.* Berlin, 1924.

[1937] *Methoden der Mathematischen Physik, II.* Berlin, 1937.

[1962] *Methods of Mathematical Physics, II*". New York, 1962.

Crowe, M. J.

[1975] Ten "laws" concerning conceptual change in mathematics. *Hist. Math.*, **2** (1975), pp. 469–470.

De Jager, E. M.

[1964] Applications of distributions in mathematical physics. *Mathematical Center Tracts, No. 10.* Mathematisch Centrum, Amsterdam, 1964.

Demidov, S. S.

[1977] Notion de solution des équations différentielles aux dérivées partielles et la discussion sur la vibration d'une corde aux XVIII siècle. *XVth International Congress of the History of Science, Edinburgh 1977.* Papers by Soviet Scientists. Section III: Mathematics and Mechanics since 1600. Moscow 1977.

Deny, J.

[1950] Les potentiels d'énergie finie. *Acta Math.*, **82** (1950), pp. 107–183.

Deny, J. and Lions, J. L.

[1953/54] Les espaces du type de Beppo Levi. *Ann. Inst. Fourier*, **5** (1953/54), pp. 305–370.

Dieudonné, J.

[1964] Recent developments in mathematics. *Amer. Math. Monthly*, **79** (1964), pp. 239–248.

[1975] Introductory remarks on algebra, topology and analysis. *Hist. Math.*, **2** (1975), pp. 537–548.

[1978] *Abrégé d'Histoire des Mathématiques 1700–1900*, Vols. I and II. Paris, 1978.

Dieudonné, J. and Schwartz, L.
[1949] La dualité dans les espaces (F) et (LF). *Ann. Inst. Fourier*, **1** (1949), pp. 61–101.

Dini, U.
[1878] *Fundamenti per la Teorica Della Funzioni di Variabili Reali.* Pisa 1878. Translated and supplemented by J. Lüroth and A. Schepp as *Grundlagen für eine Theorie der Functionen einer veränderlichen reellen Grösse.* Leipzig, 1892.

Dirac, P. A. M.
[1926] The physical interpretation of the quantum dynamics. *Proc. Roy. Soc.*, A, **113** (1926), pp. 621–641.
[1930] *ThePrinciples of Quantum Mechanics.* Oxford, 1930.
[1934] Discussion of the infinite distribution of electrons in the theory of the positron. *Proc. Camb. Phil. Soc.*, **30** (1934), pp. 150–163.
[1947] *The Principles of Quantum Mechanics*, 3rd ed. Oxford, 1947.

Dirichlet, P. G. L.
[1829] Sur la convergence des séries trigonométriques qui servent à représenter une fonction arbitraire entre des limites donnés. *J. Reine Angew. Math.*, **4** (1829), pp. 157–169. Werke I, p. 117–132.
[1837] Über die Darstellung ganz willkürlicher Funktionen durch sinus- und cosinusreihen. *Rept. der Phys.*, **1** (1837), Werke I, pp. 133–160.
[1876] *Vorlesungen über die in umgekehrten Verhältniss des Quadrats der Entfernung wirkenden Kräfte.* Edited by F. Grube. Leipzig, 1876.

Ditkin, V. A. and Prudnikov, A. P.
[1968] Operational calculus. *Prog. Math.* **1** (1968), pp. 1–74.

Du Bois-Reymond, P.
[1879] Erläuterungen zu den Anfangsgründen der Variationsrechnung. *Math. Ann.*, **15** (1879), pp. 283–314.

Doetsch, G.
[1937] *Theorie und Anwendung der Laplace-Transformation.* Berlin, 1937.
[1950] *Handbuch der Laplace-Transformation.* Basel, 1950.

Duhamel, J. M. C.
[1847a] Sur la conductibilité des corps cristallisés pour la chaleur. *Ann. de Chimie*, **21** (1847), pp. 457–476.
[1847b] Second mémoire sur la conductibilité des corps cristallisés pour la chaleur. *Comp. Rend. Acad. Sci. Paris*, **25** (1847), pp. 459–461, 707–710.

Euler, L.
[1748] Sur la vibration des cordes. *Mém. Acad. Sci. Berlin*, **4** (1748) (publ. 1750) = *Opera Omnia*, **10** (2), pp. 63–77.
[1753] Remarques sur les mémoires precedens de M. Bernoulli. *Mem. Acad. Sci. Berlin*, **9** (1753) (publ. 1755), pp. 196–222 = *Opera Omnia*, **10** (2), pp. 233–254.
[1759] De la propagation du son. *Mem. Acad. Sci. Berlin*, **15** (1759) (publ. 1766), pp. 185–209 = *Opera Omnia*, **1** (3), pp. 428–451.
[1763] De usu functionum discontinuarum in analysi. *Novi Comm. Acad. Sci. Petrop.*, **11** (1763) (publ. 1768), pp. 67–102 = *Opera Omnia*, **23** (1), pp. 74–91.
[1765a] Eclaircissement sur le mouvement des cordes vibrantes. *Micell. Tourin* **3** (1762–65) (publ. 1766), pp. 1–26 = *Opera Omnia*, **10** (2), pp. 377–396.

[1765b] Sur le mouvement d'une corde, qui au commencement n'a été ébranlée que dans une partie. *Mem. Acad. Sci. Berlin*, **21** (1765) (publ. 1767), pp. 307–334 = *Opera Omnia*, **10** (2), pp. 426–450.

[1765c] Eclaircissemens plus detailles sur la generation et la propagation et sur la formation de l'echo. *Mem. Acad. Sci. Berlin*, **21** (1765) (publ. 1767), pp. 335–363 = *Opera Omnia*, **1** (3), pp. 540–567.

Evans, G. C.

[1914] On the reduction of integro-differential equations. *Trans. Amer. Math. Soc.*, **15** (1914), pp. 477–496.

[1920] *Fundamental Points of Potential Theory.* Rice Institute Pamphlet 7 (1920), pp. 252–329.

[1927] *Logarithmic Potential—Discontinuous Dirichlet and Neumann Problems.* New York, 1927.

[1928] Note on a theorem of Bôcher. *Amer. J. Math.*, **50** (1928), pp. 123–126.

[1933] Complements of potential theory II. *Amer. J. Math.*, **55** (1933), pp. 29–49.

[1935] On potentials of positive mass I and II. *Trans. Amer. Math. Soc.*, **37** (1935), pp. 226–253; **38** (1935) pp. 201–236.

Fang, J.

[1970] *Bourbaki. Towards a Philosophy of Modern Mathematics.* New York, 1970.

Fantappié, L.

[1943a] *Teoria de Los Funcionales Analiticos y sus Aplicaciones.* Barcelona, 1943.

[1943b] L'indicatrice proiettiva dei funzionali lineari e i prodotti funzionali proiettive. *Annali dei Mat.*, IV, **22** (1943), p. 181.

Finley, M.

[1973] *Democracy Ancient and Modern.* New Brunswick, N.J., 1973.

Fischer, E.

[1907] Sur la convergence en moyenne. *Comp. Rend. Acad. Sci. Paris*, **144** (1907), pp. 1022–1024.

Fourier, J. B. J.

[1811] Théorie du mouvement de la chaleur dans les corps solides. *Mem. Acad. Sci. Paris*, **4** (1819/20, publ. 1824), pp. 185–555; *Ibid.* **5** (1821/22, publ. 1826), pp. 151–246.

[1822] *Théorie Analytique de la Chaleur.* Paris, 1822 = *Oeuvres I.* Edited by Darboux.

Fréchet, M.

[1906] Sur quelques points du calcul fonctionel. Thèse. Paris, 1906. *Rend. Circ. Mat. Palermo*, **22** (1906), pp. 1–74.

Friedrichs, K. O.

[1927] Die Randwert- und Eigenwertprobleme aus der Theorie der elastischen Platten. *Math. Ann.*, **98** (1927), pp. 205–247.

[1934] Spektraltheorie halbbeschränkter Operatoren und Anwendung auf Spektralzerlegung von Differentialoperatoren. *Math. Ann.*, **109** (1934). Teil I, pp. 465–487; Teil II, pp. 685–713.

[1939] On differential operators in Hilbert spaces. *Amer. J. Math.*, **61** (1939), pp. 523–544.

[1944] The identity of weak and strong extensions of differential operators. *Trans. Amer. Math. Soc.*, **55** (1944), pp. 132–151.

Fubini, G.

[1907] Il principio di minimo e i teoremi esistenza per i problemi di contorno relativi alle equazioni alle derivate parziali di ordine pari. *Rend. Circ. Mat. Palermo*, **23** (1907), pp. 58–84.

Grabiner, J.
[1975] The mathematician, the historian and the history of mathematics. *Hist. Math.*, **2** (1975), pp. 439–447.

Gillis, P.
[1943] Sur les formes différentielles et al formule de Stokes. *Mem. Acad. Roy. Belgique*, **20** (1943), pp. 1–95.

Goursat, E. J. B.
[1896/98] *Leçons sur l'Intégration des Equations aux Dérivées Partielles du Second Ordre*. Vol. I, Paris, 1896, Vol. II, Paris, 1898.
[1911] *Cours d'Analyse Mathématique*. Paris, 1911.

Green, G.
[1828] *An Essay on the Applicability of Mathematical Analysis to the Theories of Electricity and Magnetism*. Nottingham, 1828.

Grothendieck, A.
[1953] Sur certaines espaces de fonctions holomorphes I. *J. Reine Angew. Math.*, **192** (1953), p. 35.

Hadamard, J.
[1903] Sur les opérations fonctionnelles. *Comp. Rend. Acad. Sci. Paris*, **136** (1903), p. 351 = *Oeuvres* **1**, p. 405.
[1904/5] Recherches sur les solutions fondamentales et l'intégration des équations linéaires aux dérivées partielles. *Ann. Ec. Norm. Sup.*, **21** (3) (1904), pp. 535–556; *Ibid.*, **22** (3) (1905), pp. 101–141.
[1908] Théorie des équations aux dérivées partielles linéaires hyperboliques et du problème de Cauchy. *Acta Math.*, **31** (1908), pp. 333–380.
[1932] *Le Problème de Cauchy et les Équations aux Dérivées Partielles Linéaires Hyperboliques*. Paris, 1932.

Hahn, H.
[1924] Über Fouriersche Reihen und Integrale. Bericht über die Jahresversammlung zu Innsbruck, Sept. 1924. *Jahresber. Deut. Math. Ver.*, **33** (1925), p. 107.
[1925] Über die Methode der arithmetischen Mittel in der Theorie der verallgemeinerten Fourier'schen Integrale. *Sitz. Akad. Wiss. Wien. Math. Nat. Kl. Abt. II*, **134** (1925), pp. 449–470.
[1926] Über eine Verallgemeinerung der Fourierschen Integralformel. *Acta Math.*, **49** (1926), pp. 301–353.

Halperin, I.
[1937] Closures and adjoints of linear differential operators. *Ann. of Math.*, **38** (1937), p. 880.

Hankel, H.
[1870] Untersuchungen über die unendlich oft oszillierenden und unstetigen Funktionen. Tübingen 1870 = *Math. Ann.*, **20** (1882), pp. 63–112.

Hanson, N. R.
[1963] *The Concept of the Positron*. Cambridge, 1963.

Harnack, A.
[1882] Vereinfachung der Beweise in der Theorie der Fourierschen Reihe. *Math. Ann.*, **19** (1882), pp. 235–279.
[1887] Über die mit Ecken behafteten Schwingungen gespannter Saiten. *Math. Ann.*, **29** (1887), pp. 486–499.

Hawkins, T.
[1970] *Lebesgue's Theory of Integration*. Madison, London, 1970.

Heaviside, O.
[1893] On operators in physical mathematics. *Proc. Roy. Soc. London*, **52** (1893), pp. 504–529.
[1899] *Electromagnetic Theory*, Vol. II. London, 1899.
[1912] *Electromagnetic Theory*, Vol. III. London, 1912.

Heisenberg, W. and Pauli, W.
[1929] Zur Quantendynamik der Wellenfelder. *Z. Phys.*, **56** (1929), pp. 1–61 = *Pauli's Collected Sci. Papers II*, pp. 354–414.

von Helmholtz, H. L. F.
[1865] *Die Lehre von den Tonempfindungen*, 2nd ed. Braunschweig, 1865.

Hilbert, D.
[1900] Mathematische probleme. Vortrag gehalten auf dem internationalen Mathematiker Kongress zu Paris 1900. *Arch. Math. und Phys.*, **1** (3) (1901), pp. 44–63, 213–329 = *Ges. Abh.*, **3** (1935), pp. 290–329 = *Ostwalds Klassiker*, nr. 252, Leipzig, 1971.
[1905] Über das Dirichletsche Prinzip. *J. Reine Angew. Math.*, **129** (1905), pp. 63–67.
[1912] *Grundzüge einer allgemeinen Theorie der linearen Integralgleichungen*. Leipzig, 1912.

Hilbert, D., von Neumann, J. and Nordheim, L.
[1927] Über die Grundlagen der Quantenmechanik. *Math. Ann.*, **98** (1927), pp. 1–30 = *von Neumann Collected Works*, 1, pp. 104–133.

Hyperfunctions:
[1973] *Hyperfunctions and Theoretical Physics*. Berlin, 1975. Papers given at a Colloquium at Nice, 1973.

Ignatowsky, W.
[1909/10] *Die Vektoranalysis*. Leipzig, 1909/10.

Infeld, L. and Plebański,
[1957] On a further modification of Dirac's δ-functions. *Bull. Acad. Polon. Sci. Cl. III*, **5** (1957), no. 1.

Jammer, M.
[1966] *The Conceptual Development of Quantum Mechanics*. New York, 1966.

Jordan, P. and Pauli, W.
[1928] Zur Quantenelektrodynamik ladungsfreier Felder. *Z. Phys.*, **47** (1928), pp. 151–173 = *Pauli's Collected Sci. Papers*, II, pp. 331–354.

Josephs, H. J.
[1946] *Heaviside's Electric Circuit Theory*. New York, 1946.

Kirchoff, G.
[1882] Zur Theorie der Lichtstrahlen. *Sitz. K. Preuss. Akad. Wiss. Berlin*, (1882), p. 641–669.
[1891] *Vorlesungen über Mathematische Physik*, II. Leipzig, 1891.

Kline, M.
[1972] *Mathematical Thought from Ancient to Modern Times*. New York, 1972.
[1973] *Why Johnny Can't Add. The Failure of the New Math*. New York, 1973.

[1977] *Why the Professor Can't Teach: Mathematics and the Dilemma of University Education.* New York, 1977.

Koebe, P.
[1906] Herleitung der partiellen Differentialgleichung der Potential-funktion aus deren Integraleigenschaft. *Sitz. Berlin Math. Ges.* **5** (1906), pp. 39–42.

Koizumi, S.
[1931] On Heaviside's operational solution of a Volterra's integral equation when its nucleus is a function of $(x - \xi)$. *Phil. Mag.*, **11** (7) (1931), pp. 432–441.

Kondrachov, V.
[1938] Sur certaines Évaluations pour les familles de fonctions vérifiant quelques inégalités intégrales. *Dokl. Acad. Sci. URSS*, **18** (1938), pp. 235–240.

Koppelman, E.
[1971/72] The calculus of operations and the rise of abstract algebra. *Arch. Hist. Ex. Sci.*, **8** (1971/72), pp. 155–242.
[1975] Progress in mathematics. *Hist. Math.*, **2** (1975), pp. 457–463.

Korevaar, J.
[1955] Distributions defined from the point of view of applied mathematics. *Konikl. Ned. Akad. Wetenskap.*, ser. A, **58** (1955), pp. 368–389, 483–503, 663–674.

Krylov, V. I.
[1947] Sur l'existence des dérivées généralisées des fonctions sommables. *Dokl. Acad. Sci. URSS*, **55** (1947), pp. 375–381.

König, H.
[1953] Neue Begründung der Theorie der "Distributionen". *Math. Nachr.*, **9** (1953), pp. 129–148.

Köthe, G.
[1951] Dualität in der Funktionentheorie. *J. Reine Angew. Math.*, **191** (1953), pp. 30–53 (Eingegangen 1951).
[1952] Die Randverteilungen analytischer Funktionen. *Math. Z.*, **57** (1952/53), pp. 13–33 (Eingegangen 1952).

Lagrange, J. L.
[1759] Recherches sur la nature et la propagation du son. *Misc. Tauren.*, **1** (1759), pp. 1–112 = *Oeuvres*, 1, pp. 39–148.
[1760/61] Nouvelles recherches sur la nature et la propagation du son. *Misc. Tauren.*, **2** (1760/61) = *Oeuvres*, 1, pp. 151–332.
[1764/65] Correspondence with d'Alembert. Letters from the period September 1, 1764 to March 2, 1765. In Lagrange's *Oeuvres*, 13.

Laplace, P. S.
[1772] Memoire sur les suites. *Mem. Acad. Sci. Paris*, 1772 (publ. 1779) = *Oeuvres*, 10, pp. 1–89.
[1812] *Théorie Analytique des Probabilités.* Paris, 1812 = *Oeuvres*, 7.

Laugwitz, D.
[1959] Eine Einführung der δ Funktionen. *Münchner Ber.* (1959), pp. 41–59.
[1961] Anwendung unendlich kleiner Zahlen. I. Zur Theorie der Distributionen. *J. Reine Angew. Math.*, **207** (1961), pp. 53–60.

Laugwitz, D., and Schmieden, C.
[1958] Eine Erweiterung der Infinitesimalrechnung. *Math. Z.*, **69** (1958), pp. 1–39.

Lebesque, H.

[1902] Intégral, longueur, aire. *Ann. Mat. Pura. Appl.*, ser. 3, **7** (1902), pp. 231–359 (Lebesgue's thesis).

[1904] *Leçons sur l'Intégration et la Recherche des Fonctions Primitives.* Paris, 1904.

[1905] Sur les fonctions représentables analytiquement. *J. Math. Pures Appl.*, **1** (1905), pp. 139–216 = *Oeuvres Scientifiques*, 3, pp. 103–180.

[1906] *Leçons sur les Séries Trigonométriques.* Paris, 1906.

Leray, J.

[1934] Sur le mouvement d'un liquide visqueux emplissant l'espace. *Acta. Math.*, **63** (1934), pp. 193–248.

Levi, B.

[1906] Sul principo di Dirichlet. *Rend. Circ. Mat. Palermo*, **22** (1906), pp. 293–359.

Levinson, N., et al.

[1966] Norbert Wiener 1894–1964. *Bull. Amer. Math. Soc.*, **72**, no. 1, part II (1966).

Lewis, D. C.

[1933] Infinite systems of ordinary differential equations. *Trans. Amer. Math. Soc.*, **35** (1933), pp. 792–823.

Lévy, P.

[1926] Le calcul symbolique de Heaviside. *Bull. Sci. Math.*, **50** (2) (1926), pp. 174–92.

Lévy, P., Mandelbrojt, S., Malgrange, B., and Malliavin, P.

[1967] *La Vie et l'Oeuvre de Jacques Hadamard*, No. 16 in the series l'Enseignement Mathématiques, 1967.

Lighthill, M. J.

[1958] *Introduction to Fourier Analysis and Generalized Functions.* New York, 1958.

Ljusternik, L. A., and Visik, M. I.

[1959] Sergei Lvovich Sobolev on his 50th birthday. *Usp. Mat. Nauk*, **14** (1959), pp. 203–214. (Russian).

Lützen, J.

[1978] Funktionsbegrebets udvikling fra Euler til Dirichlet. *Nord. Mat. Tidskr.*, **25/26** (1978), pp. 5–32.

[1979] Heaviside's operational calculus and the attempts to rigorize it. *Arch. Hist. Ex. Sci.*, **21** (1979), pp. 161–200.

Luxemburg, W. A.

[1962] *Non-Standard Analysis; Lectures on A. Robinson's Theory of Infinitesimals and Infinitely Large Numbers.* Pasadena, California, 1962.

Mackey, G. W.

[1943] On convex topological linear spaces. *Proc. Nat. Acad. Sci. USA*, **29** (1943), pp. 315–319.

[1946] On convex topological linear spaces. *Trans. Amer. Math. Soc.*, **60** (1946), pp. 520–537.

Mandelbrojt, S. and Schwartz, L.

[1965] Jacques Hadamard. *Bull. Amer. Math. Soc.*, **71** (1965), pp. 107–129.

Maxwell, J. C.

[1873] *A Treatise on Electricity and Magnetism.* Oxford, 1873.

May, K. O.

[1973] *Bibliography and Research Manual of the History of Mathematics.* Toronto, 1973.

McShane, E. J.
[1933] Über die Unlösbarkeit eines einfachen Problems der Variationsrechnung. *Nachr. Ges. Wiss. Göttingen* (1933), pp. 358–364.
[1940] Generalized curves. *Duke Math. J.*, **6** (1940), pp. 513–536.

Mehrtens, H.
[1979] *Die Entstehung der Verbandstheorie*. Hildesheim, 1979.

Menger, K.
[1936] Courbes minimisantes non rectificables et champs généraux de courbes admissibles dans le calcul des variations. *Comp. Rend. Acad. Sci. Paris*, **202** (1936), pp. 1648–1650.

Mikusiński, J. G.
[1948] Sur la méthode de généralisation de Laurent Schwartz et sur la convergence faible. *Fund. Math.*, **35** (1948), pp. 235–239.
[1950] Sur les fondements du calcul opératoire. *Studia Math.*, **11** (1950), pp. 41–70.
[1959] *Operational Calculus*. Warszawa, 1959.

Mikusiński, J. G., and Sikorski, R.
[1957] *The Elementary Theory of Distributions*. Rozprawy Matematyczne, Warszawa, 1957/61.

Monge, G.
[1807] *Application de l'Analyse a la Géométrie*. Paris, 1807. 5th ed. Corrigée et annotée par M. Liouville, Paris, 1850.

Monna, A. F.
[1973] *Functional Analysis in Historical Perspective*. Utrecht, 1973.
[1975] *Dirichlet's Principle. A Mathematical Comedy of Errors and Its Influence on the Development of Analysis*. Utrecht, 1975.

Morrey, C. B., Jr.
[1933] A class of representations of manifolds, I. *Amer. J. Math.*, **55** (1933), pp. 683–707.
[1940] Functions of several variables and absolute continuity, II. *Duke Math. J.*, **6** (1940), pp. 187–215.
[1964] Multiple integrals in the calculus of variations. Colloquium Lectures given at Amherst, Mass. 1964 at the 69th Summer Meeting of the Amer. Math. Soc.

Murray, F. J.
[1935] Linear transformations between Hilbert spaces and the application of this theory to linear partial differential equations. *Trans. Amer. Math. Soc.*, **37** (1935), pp. 301–338.

Naas, J., and Schmid, H. L.
[1961] *Mathematisches Wörterbuch*. Edited by Naas and Schmid, Stuttgart, 1961.

Nečas, J.
[1967] *Les Méthodes Directes en Théorie des Equations Elliptiques*. Paris, Prague, 1967.

Neumann, C. G.
[1877] *Untersuchungen über dass Logarithmische und Newtonsche Potential*. Leipzig, 1877.

von Neumann, J.
[1927] Mathematische Begründung der Quantenmechanik. *Göttinger Nachr.*, (1927), pp. 1–57.

[1930] Allgeneine Eigenwerttheorie Hermitescher Funktional-operatoren. *Math. Ann.*, **102** (1930), pp. 49–131.

[1932] *Mathematische Grundlagen der Quantenmechanik*. Berlin, 1932.

[1935] On complete topological spaces. *Trans. Amer. Math. Soc.*, **37** (1935), pp. 1–20.

Niessen, K. F., and van der Pol, B.

[1932] Symbolic calculus. *Phil. Mag.*, **13** (1932), pp. 537–577.

Nikodym, O.

[1933a] Sur une classe de fonctions considerées dans l'étude du problème de Dirichlet. *Fund. Math.*, **21** (1933), pp. 129–150.

[1933b] Sur un théorême de M. S. Zaremba concernant les fonctions harmoniques. *J. Math. Pures Appl.*, ser. 9, **12** (1933), pp. 95–108.

[1935] Sur le principe du minimum. *Math. Cluj*, **9** (1935), pp. 110–128.

Oseen, C. W.

[1911a] Ein Satz über die Singularitäten welche in der Bewegung einer reibenden und unzusammendrückbaren Flüssigkeit auftreten können. *Arkiv. för Mat. Astron. och Fysik*, **6** (1911), no. 16.

[1911b] Über die Bedeutung der Integralgleichungen in der Theorie der Bewegung einer reibenden unzusammendrückbaren Flüssigkeit. *Archiv. för Mat. Astron. och Fysik*, **6** (1911), no. 23.

Pauli, W.

[1940] The connection between spin and statistics. *Phys. Rev.*, **58** (1940), pp. 716–722 = *Collected Sci. Papers*, Vol. 2, pp. 911–917.

Petrini, H.

[1899] Démonstration générale de l'équation de Poisson $\Delta V = -4\pi\rho$ en ne supposant que ρ soit continue. *K. Vet. Akad. Öfvers.* Stockholm, 1899.

[1908] Les dérivées premiers et secondes du potentiel. *Acta Math.*, **31** (1908), pp. 127–332.

Plancherel, M.

[1910] Contribution à l'étude de la représentation d'un fonction arbitraire par des intégrales définies. *Rend. Circ. Mat. Palermo*, **30** (1910), pp. 289–335.

[1913] Zur Konvergenztheorie der Integrale $\lim_{z=\infty} \int_a^z f(x) \cos xy\, dx$. *Math. Ann.*, **74** (1913), pp. 573–578.

[1915] Sur la convergence et sur la sommation par les moyennes de Cesaro de $\lim_{z=\infty} \int_a^z f(x) \cos xy\, dx$. *Math. Ann.*, **76** (1915), pp. 315–326.

Poisson, S. D.

[1815] Mémoire sur la théorie des ondes. *Mem. Acad. Roy. Sci. Paris*, **1** (1816), pp. 71–186. Read before the Academie, 1815.

[1821/22] Mémoire sur la théorie du magnétism. *Mem. Acad. Roy. Sci. Paris*, **5** (26) (1821/22), pp. 247–338. Read, 1824.

van der Pol, B.

[1929] On the operational solution of linear differential equations and an investigation of the properties of these solutions. *Phil. Mag.*, ser. 7, **8** (1929), pp. 861–898.

Pringsheim, A.

[1907] Über das Fouriersche Integraltheorem. *Jahresber. Deut. Math. Ver.*, 16 (1907), pp. 2–16.

[1910] Über neue Güldigkeitsbedingungen für die Fouriersche Integralformel. *Math. Ann.*, **68** (1910), pp. 367–408; Supplement *Math. Ann.*, **71** (1912), pp. 289–298.

Radó, T.
[1937] *Subharmonic Functions*. Erg. der Math. 5, Heft, 1, Berlin, 1937.

Radon, J.
[1913] Theorie und Anwendungen der absolut additiven Mengenfunktionen. *Sitz. Akad. Wiss. Wien. Math. Nat. Kl.* 122 abt. IIa (1916), pp. 1295–1438.

Ravetz, J. R.
[1961] Vibrating strings and arbitrary functions. *Logic of Personal Knowledge: Essays Presented to M. Polanyi on his 70th Birthday*. London, 1961, pp. 71–88.

Rellich, F.
[1930] Ein Satz über mittlere Konvergenz. *Nachr. Ges. Wiss. Göttingen. Math. Phys. Kl.* (1930), pp. 30–35.

de Rham, G.
[1929] Integrales multiples et analysis situs. *Comp. Rend. Acad. Sci. Paris*, **188** (1929), pp. 1651–1653.
[1936] Relations entre la topologie et la théorie des intégrales multiples. *L'Enseign. Math.*, **35** (1936), pp. 213–228.
[1950] Intégrales harmoniques et théorie des intersections. *Proc. Intern. Congress Math. 1950*, II, pp. 209–215.
[1955] *Variétés Différentiables, Formes, Courants, Formes Harmoniques*. Paris, 1955.

Riemann, B.
[1854] Über die Darstellbarkeit einer Funktion durch eine trigonometrische Reihe. *Abh. Ges. Wiss. Göttingen Math. Kl.*, **13** (1867, publ. 1868), pp. 133–52.
[1858/59] Über die Fortpflanzung ebener Luftwellen von endlicher Schwingungsweite. *Abh. Ges. Wiss. Göttingen Math. Kl.*, **18** (1858/59), pp. 43–65 = *Mathematische Werke*, pp. 145–164 and Selbstanzeige der vorstehenden Abhandlung. *Abh. Ges. Wiss. Göttingen Math. Kl.*, **19** (1859).

Riemann, B., and Weber, H.
[1919] *Die Partiellen Differentialgleichungen der Mathematischen Physik*. 6th ed. Braunschweig, 1919.

Riesz, F.
[1907] Sur les systèmes orthogonaux de fonctions. *Comp. Rend. Acad. Sci. Paris*, **144** (1907), pp. 615–19; see also pp. 734–736, 1409–1411.
[1909] Sue les opérations fonctionelles linéaires. *Comp. Rend. Acad. Sci. Paris*, **149** (1909), pp. 974–76 = *Oeuvres Complètes*, pp. 400–402.
[1910] Untersuchungen über Systeme integrierbarer Funktionen. *Math. Ann.*, **69** (1910), pp. 449–497.

Riesz, M.
[1938/40] Integrales de Riemann–Liouville et potentiels. *Acta Sci. Math. Szeged.*, **9** (1938/40), pp. 1–42.
[1949] L'intégral de Riemann–Liouville et le problème de Cauchy. *Acta Math.*, **81** (1949), pp. 1–223.

Robinson, A.
[1961] Non-standard analysis. *Proc. Nederl. Akad. Wetensch.* A, **64** (1961), pp. 432–440 = *Ind. Math.*, **23** (1961), pp. 432–440.
[1966] *Nonstandard Analysis*. Amsterdam, 1966.

Sato, M.
[1958] On a generalization of the concept of function. *Proc. Japan Acad.*, **34** (1958), pp. 126–130.
[1959/60] Theory of hyperfunctions, Part I, *J. Fac. Sci. Univ. Tokyo*, Sect. I, **8** (1959), pp. 139–193; Part II, *J. Fac. Sci. Univ. Tokyo*, Sect. I, **8** (1960), pp. 387–437.

Schapira, P.
[1970] *Théorie des Hyperfonctions*. Berlin, 1970.

Schauder, J. P.
[1935] Das Anfangswertproblem einer quasilinearen hyperbolischen Differentialgleichung zweiter Ordnung in beliebiger Anzahl von unabhängigen Veränderlichen. *Fund. Math.*, **24** (1935), pp. 213–246.

Scheeffer, L.
[1884] Allgemeine Untersuchungen über Rectification der Kurven. *Acta Math.*, **5** (1884), pp. 49–82.

Schmidt, E.
[1908] Über die Auflösung linearer Gleichungen mit unendlich vielen Unbekannten. *Rend. Circ. Mat. Palermo*, **25** (1908), pp. 53–77.

Schwartz, L.
[1943] *Etudes des Sommes d'Exponentielles Réelles*. Paris, 1943.
[1944] Sur certaines familles non fondamentales de fonctions continues. *Bull. Soc. Math. France*, **72** (1944), pp. 141–145.
[1945] Généralisation de la notion de fonction, de dérivation, de transformation de Fourier, et applications mathématiques et physiques. *Ann. Univ. Grenoble, Sect. Sci. Math. Phys.*, **21** (1945, publ. 1946), pp. 57–74.
[1947a] Théorie générale des fonctions moyenne-périodiques. *Ann. of Math.*, **48** (1947), pp. 857–929.
[1947b] Théorie des distributions et transformation de Fourier. Colloque C.N.R.S. 15 "Analyse Harmonique". Nancy 1947, pp. 1–8 (publ. 1949).
[1947/48] Théorie des distributions et transformation de Fourier. *Ann. Univ. Grenoble, Sect. Math. Phys.*, **23** (1947/48), pp. 7–24.
[1948] Généralisation de la notion de fonction et dérivation: théorie des distributions. *Ann. Telecomm.*, **3** (1948), pp. 135–140.
[1949] Les mathématiques en France pendant et après la guerre. *Proc. II Canadian Math. Congr. Vancouver (1949)*, pp. 49–67 (publ. 1951).
[1950] Théorie des noyaux. *Proc. Intern. Congr. Math. Cambridge Mass. I (1950)*, pp. 220–230.
[1950/51] *Théorie des Distributions*. Vol. I, Paris, 1950. Vol. II, Paris, 1951.
[1961] *Méthodes Mathématiques pour les Sciences Physiques*. Paris, 1961.
[1969] *Application des Distributions à l'Étude des Particules Élémentaires en Mécanique Quantique Relativiste*. Paris, 1969 (English version 1968).
[1974] *Notice sur les Travaux Scientifiques de Laurent Schwartz*. Autobiography written to the Académie des Sciences.
[1978] Interview: Information obtained in an interview of L. Schwartz at his home in Paris, December 1978.

Sikorski, R.
[1954] A Definition of the notion of distribution. *Bull. Acad. Pol. Sci.*, cl. 3, **2** (1954), pp. 207–11.

e Silva, J. S.
[1955] Sur une construction axiomatique de la théorie des distributions. *Revista Fac. Sci. Univ. Lisboa*, **4** (2) (1955), pp. 79–186.

Słowikowski, W.
[1955] On the theory of operator systems. *Bull. Acad. Pol. Sci.*, cl. 3, **3** (1955), pp. 3–6 (see also pp. 137–142).

Smith, J. J.
[1925] An analogy between pure mathematics and the operational mathematics of Heaviside by means of the theory of H-functions. *J. Franklin Inst.*, **200** (1925) pp. 519–534, 635–672, 775–814.
[1928] Heaviside's operators and contour integrals. *Atti. Congr. Intern. Mat. Bologna*, **5** (1928), pp. 309–335.

Sobolev, S. L.
[1933] Sur les vibrations d'un demi-plan et d'une couche à conditions initiales arbitraires. *Mat. Sb.*, **40** (1936), pp. 236–266.
[1934] Nouvelle méthode de résolution du problème de Cauchy pour les équations aux dérivées partielles du second ordre. *Dokl. Acad. Sci. URSS*, **1** (2) (1934), pp. 433–438 (Russian and French).
[1935a] Obshchaya teoriya difraktsii voln na rimanovykh poverkhnostyakh. *Trav. Inst. Steklov. Tr. Fiz.-mat. in-ta*, **9** (1935), pp. 39–105.
[1935b] Le problème de Cauchy dans l'espace des fonctionelles. *Dokl. Acad. Sci. URSS*, **7** (3) (1935), pp. 291–294.
[1936a] Méthode nouvelle à résoudre le problème de Cauchy pour les équations linéaires hyperboliques normales. *Mat. Sb.*, **1** (43) (1936), pp. 39–71.
[1936b] Sur quelques évaluations concernant les familles de fonctions ayant des dérivées à carré intégrable. *Dokl. Acad. Sci. URSS*, **1** (7) (1936), pp. 279–282 (corrected in *Doklady* **3** (12), p. 107).
[1936c] Problème limite fondamental pour les équations polyharmoniques dans un domaine au contour dégénéré. *Dokl. Acad. Sci. URSS*, **3** (7) (1936), pp. 311–314.
[1938a] Sur un théorème de l'analyse fonctionelle. *Dokl. Acad. Sci. URSS*, **20** (1) (1938), pp. 5–9.
[1938b] Sur un théorème de l'analyse fonctionelle. *Mat. Sb.*, **46** (4) (1938), pp. 471–496. (Russian with French summary, pp. 496–497.)
[1963] *Applications of Functional Analysis in Mathematical Physics*. Amer. Math. Soc. Transl. Math. Mono., Providence, RI, 1963.
[1964] *Partial Differential Equations of Mathematical Physics*. Oxford, 1964.

Stieltjes, T. J.
[1894] Recherches sur les fractions continues. *Ann. Fac. Sci. Toulouse*, **8** (1894), pp. 68–122 = *Oeuvres Complètes*, Vol. 2.

Stone, M.
[1932] *Linear Transformations in Hilbert Space*. New York, 1932.
[1961] The revolution in mathematics. *Amer. Math. Monthly*, **68** (1961), pp. 715–734 = *Liberal Education*, **47** (1961), pp. 304–334.

Sumpner, W. E.
[1931] Impulse functions. *Phil. Mag.*, **11** (7) (1931), pp. 345–368.

Taton, R.
[1950] Un texte inédit de Monge "Reflexions sur les équations aux différences partielles". *Osiris*, **9** (1950), pp. 44–61.
[1951] *L'Oeuvres Scientifiques de Monge*. Paris, 1951.

Temple, G.
[1953] Theories and applications of generalized functions. *J. London Math. Soc.*, **28** (1953), pp. 134–148.
[1955] The theory of generalized functions. *Proc. Roy. Soc. London*, ser. A, **228** (1955), pp. 175–190.

Thomson, W. (Lord Kelvin)
[1847] Theorems with reference to the solution of certain partial differential equations. *Cambridge, Dublin Math. J.*, **3** (1848), p. 84 = *J. Math.*, **12** (1847), p. 496.

Tillmann, H. G.
[1953] Randverteilungen analytischer Funktionen und Distributionen. *Math. Z.*, **59** (1953), pp. 61–83.
[1957] Die Fortsetzung analytischer Funktionale. *Abh. Mat. Sem. Univ. Hamburg*, (1957), pp. 139–197.
[1961a] Distributionen als Randverteilungen analytischer Funktionen II. *Math. Z.*, **76** (1961), pp. 5–21.
[1961b] Darstellung der Schwartschen Distributionen durch analytische Funktionen. *Math. Z.*, **77** (1961), pp. 106–124.

Titchmarsh, E. C.
[1926] The zeros of certain integral functions. *Proc. London Math. Soc.*, **25** (1926), pp. 283–302.

Tolhoek, H. A.
[1949] A mathematical justification of the use of the Dirac delta-function and other improper functions, with applications. Unpublished manuscript from Utrecht University 1949. Kindly sent to me from the author.
[1978] Interview: Information obtained in an interview of H. A. Tolhoek at his office in Groningen, December 1978.

Tonelli, L.
[1921] *Fundamenti di Calcolo delle Variazioni. Vols. I and II.* Bologna, 1921.
[1926a] Sulla quadratura delle superfice. *Atti Reale Accad. Lincei*, **3** (6) (1926), pp. 357–362, 445–450, 633–638.
[1926b] Sur la quadrature des surfaces. *Comp. Rend. Acad. Sci. Paris*, **182** (1926), pp. 1198–1200.
[1928/29] Sulle funzioni di due variabili assolutamente continue. *Mem. Accad. Sci. Inst. Bologna. Sci. Fis.*, **6** (8) (1928–29), pp. 81–88.
[1929] Sur la semi-conductivité des intégrales doubles du calcul des variations. *Acta Math.*, **53** (1929), pp. 325–346.

Treves, F.
[1967] *Topological Vector Spaces, Distributions and Kernels.* New York, 1967.
[1975] *Basic Linear Partial Differential Equations.* New York, 1975.

Truesdell, C. A.
[1960] Editor's introduction to Euler's *Opera Omnia*, **13** (2), pp. IX-CV. Lausanne 1960.

Ulam, S. M.
[1976] *Adventures of a Mathematician.* New York, 1976.

Weber, R. H., and Gans, R.
[1916] *Repertorium der Physik. I. Mechanik und Wärme. Bearbeitet von R. H. Weber und P. Hertz.* Berlin, 1916.

Weil, A.
[1948] L'avenir des mathématiques. In *Les Grand Courants de la Pensée Mathématique*. Edited by F. le Lionais. 1948.

Wentzel, G.
[1943] *Einführung in die Quantentheorie der Wellenfelder*. Wien, 1943.

Weyl, H.
[1913] Über die Randwertaufgabe der Strahlungstheorie und asymptotische Spectralgesetze. *J. Reine Angew. Math.*, **143** (1913), pp. 177–202.
[1940] The method of orthogonal projection in potential theory. *Duke Math. J.*, **7** (1940), pp. 411–444.
[1951] A half-century of mathematics. *Amer. Math. Monthly*, **58** (1951), pp. 523–553.

Wiener, N.
[1925] On the representation of functions by trigonometric integrals. *Math. Z.*, **24** (1925), pp. 575–616.
[1926a] The harmonic analysis of irregular motion. *J. Math. Phys.*, **5** (1926), pp. 99–121.
[1926b] The operational calculus. *Math. Ann.*, **95** (1926), pp. 557–584.
[1927] Laplacians and continuous linear functionals. *Acta Litt. Sci. Szeged*, **3** (1927), pp. 7–16.
[1930] Generalized harmonic analysis. *Acta Math.*, **55** (1930), pp. 117–258.
[1938] The historical background of harmonic analysis. Amer. Math. Soc. Semicentennial Publ. Vol. II. Semicentennial Addresses of the Amer. Math. Soc. New York, 1938.

Vitali, G.
[1904/05] Sulle funzioni integrali. *Atti. Acc. Sci. Torino*, **40** (1904/05), pp. 1021–1034.
[1907/08] Sui gruppi di punti e sulle funzioni di variabili reali. *Atti. Acc. Sci. Torino*, **43** (1907/08), pp. 229–246.

Volterra, V.
[1881] Sui principii del calcolo integrale. *Giorn. Mat.*, **19** (1881), pp. 33–72.
[1894] Sur les vibrations des corps élastiques isotropes. *Acta Math.*, **18** (1894), pp. 161–232.

Young, L. C.
[1933] On approximation by polygons in the calcülus of variations. *Proc. Roy. Soc. London* (A), **141** (1933), pp. 325–341.
[1938] Necessary conditions in the calculus of variations. *Acta Math.*, **69** (1938), pp. 229–258.

Youschkevich, A. P.
[1976] The concept of function up to the middle of the 19th century. *Arch. Hist. Ex. Sci.*, **16** (1976), pp. 37–85.

Zaremba, S.
[1909] Sur le principe du minimum. *Bull. Int. Acad. Cracovie*, **7** (1909), pp. 197–264.
[1927] Sur un problème toujours possible comprenant, a titre de cas particuliers, le problème de Dirichlet et celui de Neumann. *J. Math. Pures et Appl.*, **6** (9) (1927), pp. 127–163.

Zeilon, N.
[1911] Das Fundamentalintegral der allgeneinen partiellen Differentialgleichungen mit konstanten Koëffizienten. *Arkiv. för Mat. Astron. och Fysik*, **6** (1911), no. 38.

PROBLEMS Chart 1

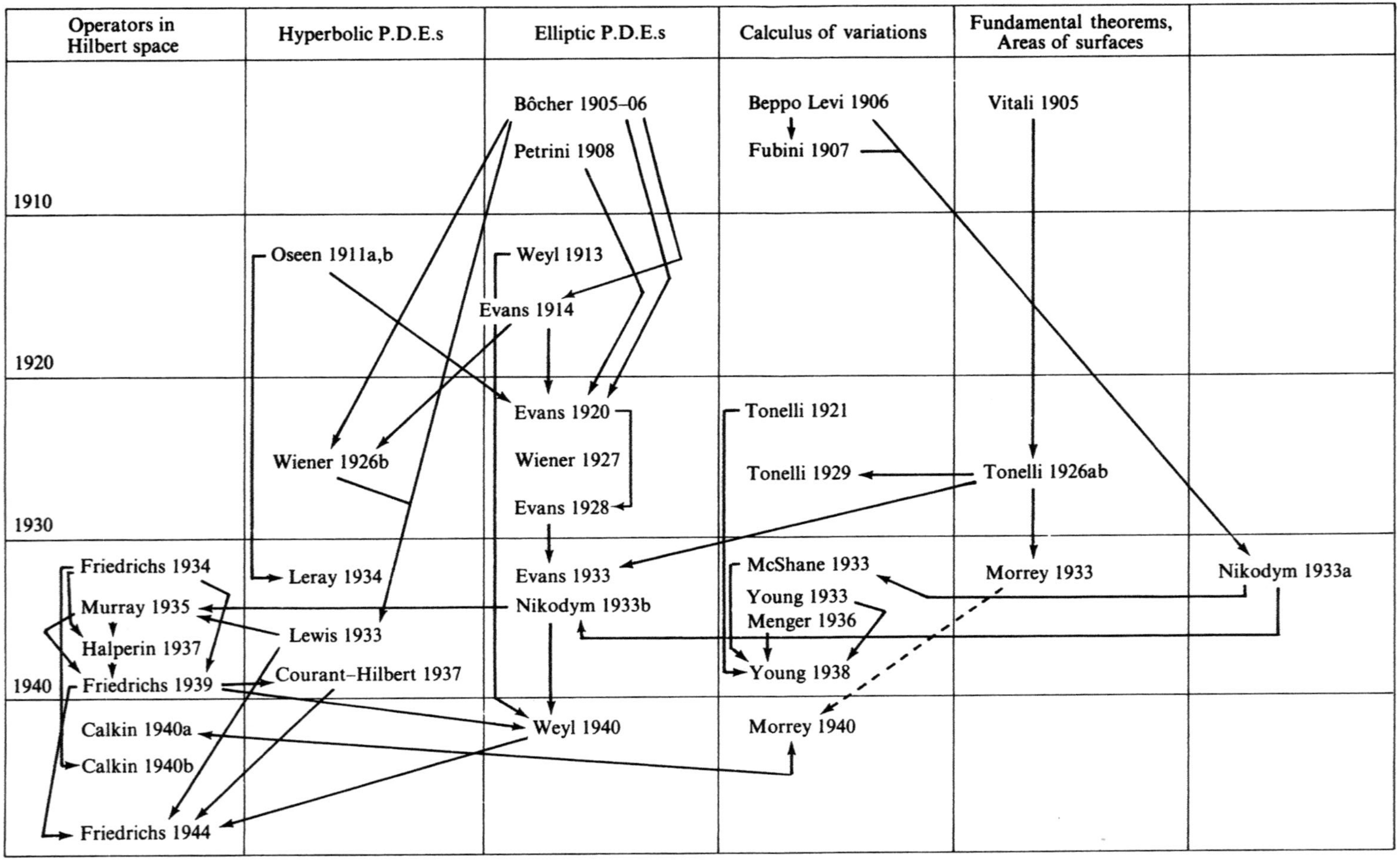

GENERALIZATION METHODS Chart 2

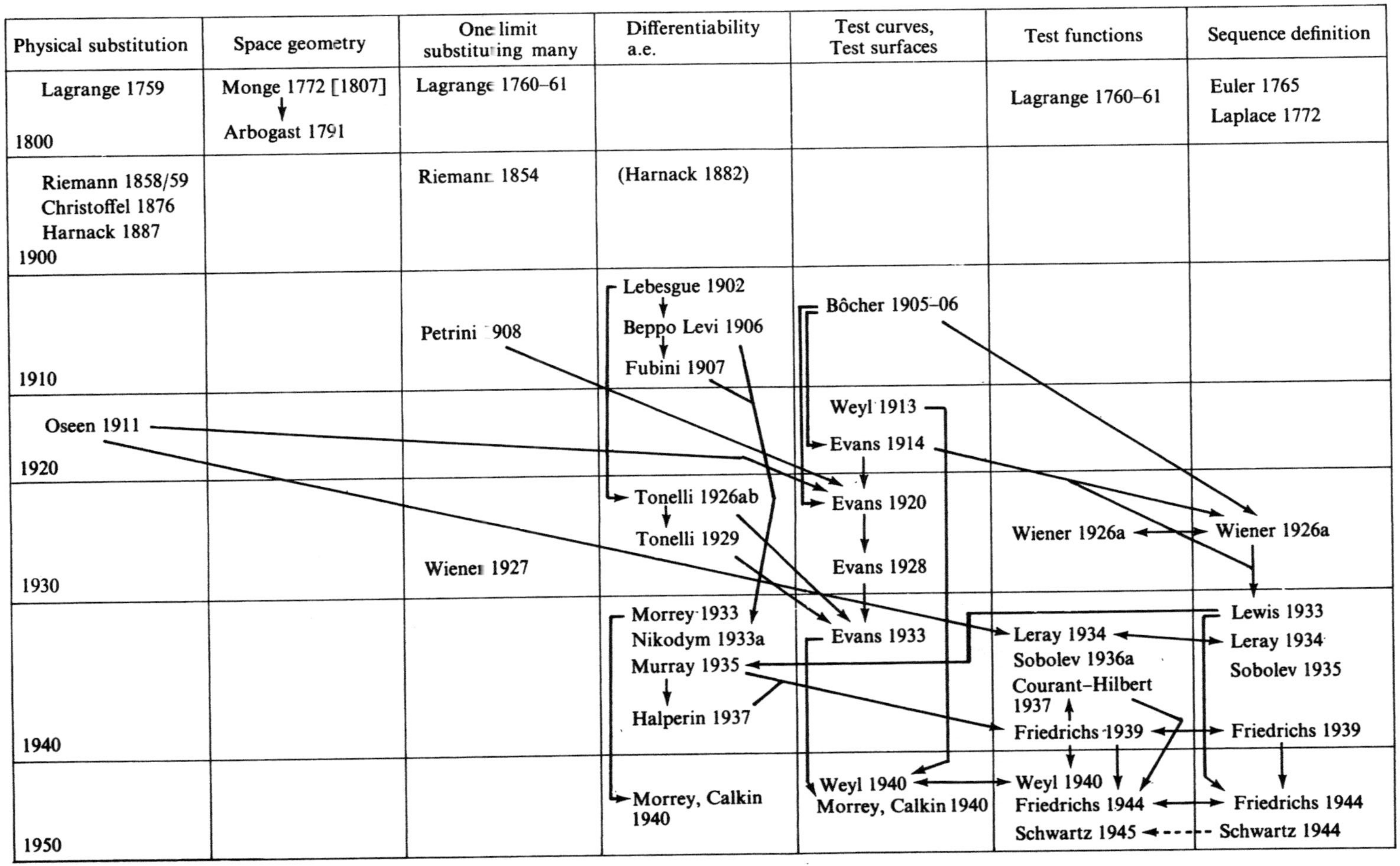

Index

GPSR Compliance

The European Union's (EU) General Product Safety Regulation (GPSR) is a set of rules that requires consumer products to be safe and our obligations to ensure this.

If you have any concerns about our products, you can contact us on ProductSafety@springernature.com

In case Publisher is established outside the EU, the EU authorized representative is:

Springer Nature Customer Service Center GmbH
Europaplatz 3
69115 Heidelberg, Germany

Batch number: 10682317

Printed by Printforce, the Netherlands